Chicago Public Library
C
P
L
REFERENCE
Form 178 rev. 1-94

D0966360

THE NATIONAL ACADEMIES

National Academy of Sciences
National Academy of Engineering
Institute of Medicine
National Research Council

The **National Academy of Sciences** is a private, nonprofit, self-perpetuating society of distinguished scholars engaged in scientific and engineering research, dedicated to the furtherance of science and technology and to their use for the general welfare. Upon the authority of the charter granted to it by the Congress in 1863, the Academy has a mandate that requires it to advise the federal government on scientific and technical matters. Dr. Bruce M. Alberts is president of the National Academy of Sciences.

The **National Academy of Engineering** was established in 1964, under the charter of the National Academy of Sciences, as a parallel organization of outstanding engineers. It is autonomous in its administration and in the selection of its members, sharing with the National Academy of Sciences the responsibility for advising the federal government. The National Academy of Engineering also sponsors engineering programs aimed at meeting national needs, encourages education and research, and recognizes the superior achievements of engineers. Dr. William A. Wulf is president of the National Academy of Engineering.

The **Institute of Medicine** was established in 1970 by the National Academy of Sciences to secure the services of eminent members of appropriate professions in the examination of policy matters pertaining to the health of the public. The Institute acts under the responsibility given to the National Academy of Sciences by its congressional charter to be an adviser to the federal government and, upon its own initiative, to identify issues of medical care, research, and education. Dr. Kenneth I. Shine is president of the Institute of Medicine.

The **National Research Council** was organized by the National Academy of Sciences in 1916 to associate the broad community of science and technology with the Academy's purposes of furthering knowledge and advising the federal government. Functioning in accordance with general policies determined by the Academy, the Council has become the principal operating agency of both the National Academy of Sciences and the National Academy of Engineering in providing services to the government, the public, and the scientific and engineering communities. The Council is administered jointly by both Academies and the Institute of Medicine. Dr. Bruce M. Alberts and Dr. William A. Wulf are chairman and vice chairman, respectively, of the National Research Council.

COMMITTEE TO EVALUATE INDICATORS FOR MONITORING AQUATIC AND TERRESTRIAL ENVIRONMENTS

GORDON H. ORIANS (*Chair*), University of Washington, Seattle
MARTIN ALEXANDER, Cornell University, Ithaca, New York
PATRICK L. BREZONIK, University of Minnesota, St. Paul
GRACE BRUSH, The Johns Hopkins University, Baltimore, Maryland
EVILLE GORHAM, University of Minnesota, Minneapolis
ANTHONY JANETOS, World Resources Institute, Washington, D.C.
ARTHUR H. JOHNSON, University of Pennsylvania, Philadelphia
DANIEL V. MARKOWITZ, Malcolm Pirnie, Akron, Ohio
STEPHEN W. PACALA, Princeton University, Princeton, New Jersey
JOHN PASTOR, University of Minnesota, Duluth
GARY W. PETERSEN, Pennsylvania State University, University Park
JAMES R. PRATT, Portland State University, Portland, Oregon
TERRY ROOT, The University of Michigan, Ann Arbor
MICHAEL L. ROSENZWEIG, University of Arizona, Tucson
MILTON RUSSELL, University of Tennessee, Knoxville
SUSAN STAFFORD, Colorado State University, Fort Collins

National Research Council Staff

DAVID POLICANSKY, Project Director
JEFFREY W. JACOBS, Senior Staff Officer
ANITA A. HALL, Project Assistant
STEPHANIE L. VANN, Project Assistant

WATER SCIENCE AND TECHNOLOGY BOARD

Preface

The U.S. Environmental Protection Agency (EPA) is reevaluating its approach to environmental monitoring. In this context, the agency asked the National Research Council (NRC) to conduct a critical scientific evaluation of indicators to monitor ecological changes from either natural or anthropogenic causes. Specifically, EPA asked the NRC to identify criteria for evaluating biological indicators, to evaluate methods of indicator development, to provide examples of indicators that have proven useful, and to identify areas where further research is likely to yield more useful and powerful indicators. The NRC was also asked to examine what aspects of environmental conditions and trends should be monitored. It was also requested to identify aspects of ecosystem structure and functioning that have been particularly difficult to characterize by means of indicators and to assess whether new approaches might allow some of these aspects to be better characterized.

In response to EPA's request, the NRC established the Committee to Evaluate Indicators for Monitoring Aquatic and Terrestrial Environments (hereafter referred to as the "Indicators Committee"). This committee has focused on ecological indicators for EPA's Environmental Monitoring and Assessment Program (EMAP), and for the needs of the ecological monitoring programs of other federal and state agencies as well. The Indicators Committee's focus on ecological indicators of environmental status and trends is timely, because the development of ecological indicators has been more difficult than that of physical indicators, which has a rich and extensive history. Today, the use of a number of physical indicators, such as atmospheric concentrations of carbon dioxide, already command considerable attention within the United States and internationally.

Ecological processes and products are varied and often complex, and large numbers of ecological indicators have been developed and used. These indicators are often intended to inform decision makers about the status and trends in populations of particular species and small groups of species, and in particular ecosystems. Useful though some of these indicators may be, they do not provide a basis for evaluating in general terms the state of the nation's ecosystems and how they are changing. Thus, most current ecological indicators have limited use for guiding national environmental policies. With this in mind, the Indicators Committee decided that its main task was to identify and characterize general ecological indicators capable of informing the public and decision makers about the overall state of the nation's ecosystems and how those ecosystems may be changing due to anthropogenic and other pressures.

Decision makers and the public need accurate information on ecological conditions and changes for three major reasons. First, a long-term record of conditions is needed as a reference to evaluate current conditions and trends. Second, detailed information on the ecological effects of various human activities and natural events—such as pollution, development, agriculture, climate change, and geomorphological events—is essential for selecting and implementing management options to address problems successfully. Finally, long-term ecological data are needed for society to measure the effectiveness and efficiency of management interventions and to improve them.

In responding to its charge, the Indicators Committee examined previous relevant reports of the National Research Council,[1] including reviews of EPA's Environmental Monitoring and Assessment Program (*Review of EPA's Environmental Management and Assessment Program: Forests and Estuaries* {1994}, *Surface Waters* {1994}, *Overall Evaluation* {1995}), *A Review of the USGS. National Water Quality Assessment Pilot Program* (1990), *Review of the Department of the Interior's Biomonitoring of Environmental Status and Trends Program: The Draft Detailed Plan* (1995), *Biologic Markers of Air-Pollution Stress and Damage in Forests* (1989), *Managing Troubled Waters: The Role of Marine Environmental Monitoring* (1990), *Animals as Sentinels of Environmental Health Hazards* (1991), and four reviews of the Minerals Management Service's Outer Continental Shelf Environmental Studies Program (*Assessment of the U.S. Outer Continental Shelf Environmental Studies Program: I. Physical Oceanography* {1990}, *II. Ecology* {1992}, *III. Social*

---

[1]All published by National Academy Press, Washington, D.C.

and Economic Studies {1992}, and *IV. Lessons and Opportunities* {1993}). Recognizing that much effort had already been expended by many capable and informed people to devise ecological indicators, the Indicators Committee also reviewed a wide array of documents produced by federal, state, and private agencies and individual investigators, many of which have been published in peer-reviewed journals and books.

The Indicators Committee gratefully acknowledges the valuable presentations made at its meetings by Robert Huggett, Vice President for Research and Graduate Studies, Michigan State University (then Assistant Administrator for Research and Development, U.S. Environmental Protection Agency); Steven Paulsen, Director of the Environmental Monitoring and Assessment Program (EMAP), National Health and Environmental Effects Research Laboratory (NHEERL), Western Ecology Division, U.S. Environmental Protection agency; Gilman Veith, Associate Director for Ecology, NHEERL, U.S. Environmental Protection Agency; Ray Wilhour, Associate Director for Science, Gulf Ecology Division, NHEERL, U.S. Environmental Protection Agency.

As always, the work of the committee has depended greatly on its supporting NRC staff. Anita Hall and Stephanie Vann have been responsible for the complex logistics involved in committee meetings, and Jeffrey Jacobs helped locate essential literature. Jim Reisa's insights and experience with the topic much improved the clarity of this report. The committee particularly acknowledges the extensive efforts and intellectual contributions of the project director, David Policansky. Carole Rosenzweig helped arranged the venue for a very productive meeting of the committee in Arizona in February 1998.

Gordon Orians, *Chair*

Acknowledgment of Reviewers

This report has been reviewed by individuals chosen for their diverse perspectives and technical expertise, in accordance with procedures approved by the National Research Council's Report Review Committee. The purpose of this independent review is to provide candid and critical comments that will assist the institution in making the published report as sound as possible and to ensure that the report meets institutional standards for objectivity, evidence, and responsiveness to the study charge. The review comments and draft manuscripts remain confidential to protect the integrity of the deliberative process. We wish to thank the following individuals for their participation in the review of this report:

Ann Bartuska, U.S. Forest Service
William C. Clark, Harvard University
J. Clarence Davies, Resources for the Future
Richard Fisher, Texas A&M University
Lisa Graumlich, Montana State University
Mark A. Harwell, University of Miami
H. Ronald Pulliam, University of Georgia
Stephen Running, University of Montana

While the individuals listed above provided constructive comments and suggestions, it must be emphasized that responsibility for the final content of this report rests entirely with the authoring committee and the institution.

Contents

ECOLOGICAL INDICATORS FOR THE NATION

Executive Summary

Indicators are designed to inform us quickly and easily about something of interest. They communicate information about conditions, and, over time, about changes and trends. Like economic indicators, environmental indicators are needed because it is not possible to measure everything.

Developing indicators and monitoring them over time can help to determine whether problems are developing, whether any action is desirable or necessary, what action might yield the best results, and how successful past actions have been. To develop and implement sound environmental policies, data are needed that capture the essence of the dynamics of environmental systems and changes in their functioning. These kinds of data then need to be incorporated into indicators.

Although no current indicators of environmental conditions or trends have the stature of the most influential economic indicators, some environmental indicators, such as global mean temperature, sea surface temperatures, and atmospheric carbon dioxide concentrations, are attracting considerable attention. Developing indicators of comparable power for ecological processes will help focus appropriate attention on ecological conditions, providing clues that could help guide significant and informed policy choices. Ecological indicators are also needed as yardsticks to measure the need for and performance of public policies and programs.

During recent decades, a variety of efforts have laid important groundwork for the development of national-level indicators to inform major policy decisions. Our work builds on these efforts.

This report has several goals: (1) to suggest criteria for selecting useful ecological indicators, (2) to provide methods for integrating complex ecological information into indicators that summarize in simple but powerful ways conditions and changes in important ecological processes and products, (3) to propose indicators that meet the suggested criteria, (4) to identify sources of data that can be used to design and compute the numerical value of indicators, and (5) to offer guidance for gathering, storing, interpreting, and communicating information from ecological monitoring.

This report concentrates on ecological indicators; this was the charge to the committee and an area in which better indicators are urgently needed. Our report does not cover other important types of environmental indicators, such as physical and chemical indicators of climate change, ozone depletion, acid precipitation, or air and water quality, although those indicators are no less important than the ones to which we give our primary attention here.

SCALES AND APPLICABILITY OF INDICATORS

Indicators can be useful at many levels—community, state, ecoregional, watershed, national, and international—and better indicators are needed at all such scales. In addition, better ways are needed of matching the scales at which indicators are useful to the scales of ecological processes. In this report, we concentrate on indicators that can support national decision making, but we also show how the methods we recommend can be used to develop indicators whose primary use would be at local and regional scales.

National ecological indicators are difficult to develop for a country as large and diverse as the United States. Determining appropriate and useful ways to aggregate information collected at small scales into indicators covering the entire country is challenging. Indeed, this problem has not yet been fully solved for all of the indicators we propose. Some will require further development and pilot studies before they are fully implemented to better understand how they respond to temporal and spatial variability. Yet national-level indicators are needed because many environmental policies are made or implemented nationally, and many international agreements need national-level information to help establish international standards. For all these reasons, the committee has focused most of its efforts on indicators that are potentially useful at a national level.

Ecological indicators that describe the state of the nation's ecosystems and command credibility and attention from the public and decision makers have been elusive. In part, this results from the complexity of ecological systems, but more attention should be given to the criteria for developing and using successful national ecological indicators. In addition, many current ecological indicators are applicable to only limited areas, to one type of ecosystem, or to the populations of one or a few species. They are useful for their intended purposes, but they cannot serve as nationwide indicators.

Some indicators have been less useful than hoped because the measures used are not clearly linked to underlying ecological processes. As a result, it has been difficult for scientists to interpret changes in those indicators. In other cases, data requirements are so complex and extensive that the indicators would be too expensive to use. Another difficulty is the frequent need to combine very different kinds of variables into a single indicator. These types of problems in indicator development and interpretation have plagued scientists and managers for many years.

CRITERIA FOR EVALUATING INDICATORS

To avoid the above pitfalls and to provide a common framework for indicators, the committee developed a general checklist of criteria for evaluating them. The checklist can be used to assess the potential importance of a proposed indicator, its properties, its domain of applicability, and its limitations, and thus how the indicator might be used. The items in this checklist follow.

- *General Importance.* Does the indicator provide information about changes in important ecological and biogeochemical processes? Does the indicator tell us something about major environmental changes that affect wide areas?

- *Conceptual Basis.* Is the indicator based on a well-understood and generally accepted conceptual model of the system to which it is applied? Is it based on well-established scientific principles? The conceptual model provides the rationale for the indicator, suggests how it should be computed, and enables us to understand the features of the indicator and how they change.

- *Reliability.* What experience or other evidence demonstrates the indicator's reliability? The best evidence for the reliability of an indicator is, of course, successful previous use. Nevertheless, all existing indicators should be analyzed retrospectively before assuming that their use should be continued. An indicator that is newly proposed inevitably lacks a historical record of reliability. Nonetheless, if it is based on a well-established scientific theory, and if a retrospective analysis has indicated that it probably would have informed us about important changes in an environmental process or product of concern, its reliability is provisionally established. Some of the indicators we have proposed are new, and like other new indicators, development and experience will be needed to make them operational.

- *Temporal and Spatial Scales.* Does the indicator inform us about national, regional, or local ecological conditions, processes, and products? Are the changes measured by the indicator likely to be short-term or long-term? Can the indicator detect changes at appropriate temporal and spatial scales without being overwhelmed by variability? To determine what an indicator indicates, the kinds of data needed to compute it, and how changes in it should be interpreted, the temporal and spatial scales of the processes measured by the indicator need to be clear.

- *Statistical Properties.* In the areas of accuracy, sensitivity, precision, and robustness, has the indicator been shown to serve its intended purpose? Is the indicator sensitive enough to detect important changes but not so sensitive that signals are masked by natural variability? Are its statistical properties understood well enough that changes in its values will have clear and unambiguous meaning?

- *Data Requirements.* How much and what kinds of information are necessary to permit reliable estimates of the indicator to be calculated? How many and what kinds of data are required for the indicator to detect a trend? Most ecological indicators depend on data gathered by means of long-term monitoring. The challenge is deciding which rates of change to watch, and to determine which of the changes observed represent significant departures from expected natural variability. Once an indicator is selected, monitoring must be used to gain experience with the likely meaning of changes in the indicator's values. Experimental studies—themselves requiring monitoring—should be used to determine whether the stress/response relationships suggested by the monitoring program are indeed causal. The use of the indicator may change as additional insights are gained into its behavior and the underlying processes that cause it to change.

- *Skills Required.* What technical and conceptual skills must the collectors of data for an indicator possess? Does the collection of input data require highly technical, specialized knowledge if the data are to be accurate, or is data collection a relatively straightforward process? An indicator capable of commanding broad attention must be based on data that are accurate and, equally important, perceived by all to be accurate.

Because the collection of data for ecological indicators (i.e., monitoring) is sometimes perceived by scientists as boring or less interesting and prestigious than "scientific research" (i.e., hypothesis-driven investigation), it is important to provide incentives for consistent and accurate data collection. One way to do that is to design monitoring programs so that the information also has scientific value (i.e., can be used to help to answer research questions). The indicators we have proposed embody hypotheses about the functioning of ecosystems. To the degree that such hypotheses can be made explicit in the design of indicators, their development and the subsequent monitoring of them should generate a great deal of valuable scientific information.

- *Data Quality.* No indicator of environmental quality is reliable unless the underlying data that are used to construct or calculate it are accurate. Attention to data quality during the archiving and computational phases cannot substitute for the quality of the input data. In this critical sense, the ultimate responsibility for data quality must lie with the investigators who collect them. Clear documentation of sampling and analytical methods is necessary if future investigators are to understand exactly how each indicator was calculated. This requirement is particularly important as methods and instrumentation change, so that data from early parts of the time series are quantifiably comparable to data from later parts of the same time series.

- *Data Archiving.* A monitoring system to track ecological indicators requires archiving capabilities that provide interested parties access to the data. For indicators that are direct representations of environmental samples, the archive simply needs to save a record of the measurements. In general, the minimum number of physical samples saved should ensure the ability to recalibrate the entire data set, should that become necessary because of changes in sampling or analytical technologies. The costs of preserving physical samples in forms that do not decay or otherwise change must be weighed against the opportunity cost of not being able to recalibrate a data set with improved or modified measurement techniques. The complete description and availability of the models and the data used to calculate indicators are just as important as the availability of the underlying data themselves; otherwise, future comparisons might actually not compare the same things. The archive must be robust enough to ensure that the time series of the indicator can be reprocessed as models improve.

- *Robustness.* For our purposes here, we define robustness in a nonstatistical sense, as an indicator's ability to yield reliable and useful numbers in the face of external perturbations. In other words, is the indicator relatively insensitive to expected sources of interference? Are technological changes likely to render the indicator irrelevant or of limited value? Can time series of measurements be continued in compatible form when measurement technologies change? To continue to gather data by outdated methods is undesirable. Nevertheless, because long-term data sets are essential for detecting most environmental trends, technological changes must be incorporated into monitoring programs in ways that do not destroy the continuity of the data sets or render consistent interpretation of the changes impossible. As pointed out in Chapter 2, cross-calibration of measurements is especially important for remotely sensed data.

- *International Compatibility.* Is the indicator compatible with indicators being developed by other nations and international groups? Not all indicators used in the United States, especially those relating to specific regions, ecosystems, or species, need to be compatible with indicators developed and used in other nations. However, national-level indicators signal changes that are likely to transcend national boundaries. Effective responses to these changes may require international action. If the signals that trigger actions are not meaningful to the affected nations, appropriate multinational responses are certain to be more difficult to mount.

- *Costs, Benefits, and Cost-Effectiveness.* Costs and benefits associated with implementing proposed ecological indicators are important because resources for monitoring are limited and should be used efficiently. The costs of developing and monitoring an indicator, which can continue to accrue as the indicator is used and refined and as new data and technologies develop, can be estimated objectively. The benefits—the value of the information obtained—are more difficult to estimate. The greater the benefits of an indicator, the higher the costs that can be justified in developing and implementing it. Cost-effectiveness is also an important criterion. If one assumes that the information an indicator yields is essential, can it be obtained for less cost in another way? If so, the indicator is not cost-effective. The value of the information was the committee's first consideration in every indicator we recommend.

THE COMMITTEE'S CONCEPTUAL MODEL FOR CHOOSING INDICATORS

To guide its selection of ecological indicators, the committee used the above criteria and a conceptual model of the factors that most strongly influence ecosystem functioning, described in Chapter 2. The goods and services that ecosystems provide to humans depend directly or indirectly on ecosystems' productivity, i.e., their ability to capture solar energy and store it as carbon-based molecules. Productivity is strongly influenced by temperature, moisture, soil fertility, and the structure and composition of ecological communities. Measures of the presence of native and exotic species are also important inputs to national ecological indicators. How the committee used this conceptual model is described in Chapter 4, where we recommend a set of indicators of the key factors that influence ecosystem functioning.

POLICY PERSPECTIVES

For use in policy making, indicators need to be understandable, quantifiable, and broadly applicable. They should provide information about important ecological processes. Indicators are more likely to be influential if there are relatively few of them and if they convey information in a form that the public and policy makers understand. It is crucially important for public confidence in an indicator that its numerical values be independent of who does the calculating: the rules for calculating an indicator from measurement data must be objective and clear. The indicators need to be credible, and therefore the people and organizations that produce them need to be credible. This recommendation is especially critical if the indicators recommended here are used as input for reporting on the status of the nation's ecosystems, as we hope they will be.

THE RECOMMENDED INDICATORS

Based on consideration of the desirable characteristics of indicators, the sources of data that underlie them, the models that support them, the criteria summarized above, and the conceptual model we used, the committee recommends the following national ecological indicators in three categories:

- As indicators of the extent and status of the nation's ecosystems, the committee recommends *land cover* and *land use*.
- As indicators of the nation's ecological capital, the committee recommends *total species diversity*, *native species diversity*, *nutrient runoff*, and *soil organic matter*.
- As indicators of ecological functioning or performance, the committee recommends *carbon storage*, *production capacity*, *net primary production*, *lake trophic status*, *stream oxygen*, and for agricultural ecosystems, *nutrient-use efficiency* and *nutrient balance*.

For each indicator recommended in this report, information is provided on the following points insofar as possible:

- Why the indicator is useful.
- The ecological model that underlies the indicator.
- The range of values the indicator can take and what the values mean.
- The temporal and spatial scales over which the indicator is likely to change.
- Whether the needed input data are already being gathered, and, if so, by whom.
- If the needed data are not being gathered, what new data are needed and who should collect them.
- The probable effects of new technologies on our ability to make the required measurements and how soon significant technological changes are likely.

In some cases, noted for each indicator, some experience will need to be gained on details of its behavior, but all the indicators are based on soundly established scientific principles and experience. The proposed indicators are in general applicable to both managed (e.g., agricultural and silvicultural) and unmanaged ecosystems; the indicators of nutrient-use efficiency and overall nutrient balance are specific to agricultural ecosystems.

Indicators of Ecosystem Extent and Status

The largest ecological changes caused by humans result from land use. The changes include replacing native biological communities with agricultural systems, changing hydrological and biogeochemical cycles, changing the Earth's surface by creating buildings and transportation corridors, and so on. These changes affect the ability of ecosystems to provide the goods and services that society depends on. For this reason, it is necessary to know about land cover and land use. In addition, an indicator of land cover provides a rough inventory of the nation's biological capital, an essential part of any suite of indicators. Information on land cover provides a reference point and is needed to calculate several other proposed indicators. This information provides a standard against which to detect and measure changes. The information and technology to calculate land cover are currently available; land use is in some ways more informative, but considerable synthesis of existing information and some new information will be required to develop that indicator.

The committee recommends a **land cover** indicator that includes aquatic and dryland ecosystems. It records the percentage of land in each of many land cover categories. Each time land cover is computed, the proportions in each category should be compared with those at the previous recording time. Data must be entered and stored separately for many categories of land cover types. Because the proportion of land in each category changes relatively slowly, land cover needs to be reported only every five years, but its values should be computed annually so it can be used as inputs to other indicators.

The land cover indicator measures the proportion of the landscape occupied by each member of a set of land cover types that comprise the total area of the nation. The major questions concern how many land cover types to recognize, how to account for their spatial configurations, and how to accommodate changes in the number and kinds of categories recognized. Changes in the proportional representation of various land cover types is the variable of interest. Satellite imagery can identify many categories of land cover.

The classification of land cover types must be comprehensive to serve other indicators that are derived from it. The number of classes of land cover types that will be needed will likely be different for each indicator, and therefore the input data should be archived at the most highly resolved and disaggregated levels.

Some indicators of land cover are currently used on less than a national scale, such as the U.S. Department of Agriculture's National Resources Inventory. The inventory covers 800,000 sites on private lands and has provided valuable information on ecological extent and condition. A national indicator that could be applied to all U.S. lands and even beyond would be even more useful.

When sufficient information is developed on land use, the committee recommends that a similar **land use** indicator be developed.

Indicators of Ecological Capital

The capacity of ecosystems to provide goods and services depends on the natural capital, both biotic and abiotic, that constitutes them. The essential capital includes physical components such as soil condition, as well as the species that drive and maintain ecosystem processes. Therefore the committee recommends indicators of species diversity, soil condition, and nutrient runoff.

The United States has affirmed many times through law, policy, and action its commitment to preserve its biological resources. Because loss of a species is irreversible, species richness is especially important to monitor. Ecological capital includes the number of species still present in the country relative to their number at the time of European contact, their distribution in today's natural and human-modified environments, and the number of species present today that are nonnative.

The first recommended indicator of ecological capital, **total species diversity**, measures the ecological capital actually present. It combines a measure of diversity with information about land cover. To avoid discounting rare species, the diversity measure is species richness, which is not weighted by population abundances. The diversity measure, based on decades' worth of experience with species-area curves for many taxa, is described in detail in Chapter 4. The indicator will be based on the land cover indicator: a diversity score will be assigned to each category of land cover based on its contribution to total species diversity. The average for the whole country—the national indicator of total species diversity—is computed by multiplying each diversity score by the total area in its land cover category, summing scores, and dividing the total by the area of the nation. This indicator of species richness should be calculated for a few representative taxa that are reasonably well known and easy to sample. The indicator's chief value is in providing a measure of total species richness. It can reflect human impacts, especially severe ones, and it also reflects many other environmental variations. It thus allows one to compare the species richness in various land cover types as well as the effects on species richness of various natural environmental and human-caused changes.

The second indicator, **native species diversity**, reflects human impact on the land. Land that has been so transformed by people that it cannot support native species that would otherwise be there carries a heavy burden caused by human activities. Thus, the indicator compares the number of native species an area of land supports with the number of native species one would expect such a landscape type to support. Total species diversity includes all the species present, both native and nonnative. Native species diversity includes only natives, because its purpose is to reflect human impacts. If humans cause a native species to be replaced by a nonnative one, native species diversity will change, indicating an impact even though total species diversity does not change. Both indicators depend similarly on land cover (or land use) for their calculation. Although lack of adequate information on many taxa will make developing these indicators difficult, the work—including developing better information about species diversity of many taxa—should be started now. Doing so will provide an incentive to learn about taxa that are not well known at present, and enough is known about some taxa to be useful now.

Soil is the source of nutrients and energy for soil biota, and soil structure influences its capacity for water retention, its susceptibility to erosion, and the fate of pollutants such as pesticides. The best indicator of soil condition is **soil organic matter** (SOM). SOM strongly influences all of these processes. Concentrations of SOM generally range from 1 to 10 percent, and it can recover or be maintained through careful management, just as it can decline through inappropriate management. Thus it is a useful indicator of agricultural soil condition as well as of the condition of unmanaged soils.

The committee recommends an indicator of **nutrient runoff**, which measures the loss of essential nutrients from the soil and is related to soil erosion, because excess nutrients, especially nitrogen and phosphorus, reduce water clarity, increase nuisance algal blooms, and increase the incidence of hypoxia (low oxygen) in waters. The adverse effects are seen in fresh waters, estuaries, and coastal waters. For these reasons, the committee recommends nutrient runoff, measured by total nitrogen and phosphorus, as an indicator of water quality. The indicator can take values from 0 (no discharge) to thousands of kilograms per square kilometer per year, with lower values being more desirable for most purposes. Because nitrogen and phosphorus runoff is largely a result of human activities, it can also be an indicator of the need for and effectiveness of environmental management.

Indicators of Ecosystem Functioning

Changes in ecosystems' productivity are usually accompanied by changes in their ability to provide goods and services important to humans. Usually, declines in productivity are undesirable, but in freshwater ecosystems, increases in productivity associated with eutrophication can be undesirable. The committee recommends three indicators of terrestrial productivity and two aquatic indicators.

Energy in the form of light is captured by chlorophyll and converted to chemical energy in the form of carbon, a process called primary production, the basis of terrestrial productivity. The committee recommends an indicator of **production capacity,** measured by total chlorophyll per unit area. It provides a direct measure of the energy-capturing capacity of terrestrial ecosystems. An equivalent measure for lakes would be total chlorophyll per unit volume. Total chlorophyll is an excellent indicator because it is strongly correlated with an ecosystem's actual capacity to capture energy. The chlorophyll per unit area ranges from 2.8 g/m^2 in tropical forests to 0.5 g/m^2 in tundra and desert ecosystems.

Next, the committee recommends an indicator of **net primary production (NPP)**, which is a direct measure of the amount of energy and carbon that has been brought into an ecosystem; it also is a measure of productivity as understood in forestry and agriculture, i.e., the amount of plant material produced in an area per year. Values of NPP range from

1,400 $g/m^2/year$ in tropical rainforests to 50 $g/m^2/year$ in American deserts. In some agricultural systems, it can reach 6,000 $g/m^2/year$.

The committee also recommends an indicator of **carbon storage,** a direct measure of the amount of carbon sequestered or released by ecosystems. It is the difference between the sum of all nonplant respiration in an ecosystem—all the CO_2 carbon produced by detritivores and animals—and net primary production. It measures the change in the total amount of carbon in an ecosystem, and hence indicates the ecosystem's carbon balance. This indicator is important in light of concerns about greenhouse gas emissions because the carbon released by a region equals the region's fossil-fuel emissions minus its ecosystem carbon storage.

A fourth indicator, **stream oxygen**, is recommended by the committee as an indicator of the ecological functioning of flowing-water ecosystems. It captures the balance between instream primary production and respiration. High stream oxygen indicates much photosynthetic activity and the likelihood of high nutrient concentrations, algal blooms, and rapid growth of leafy aquatic plants. Low stream oxygen indicates higher respiration than photosynthesis and the likelihood of organic enrichment from wastewater or high plant production upstream. Low stream oxygen usually indicates that the water is not suitable for many species of aquatic animals, including fish.

Finally, the committee recommends **trophic status of lakes** as an indicator of aquatic productivity. Such an indicator can be developed from a few key characteristics that determine the functional properties of lakes and their ability to provide the many goods and services valued by society. The key characteristics—nutrient status, net biological production, and water clarity— are closely interrelated and they are influenced by management of fertilizers, sewage, and other nutrient sources. Net biological production and water clarity can be measured by satellite imagery as well as by ground-based methods. Together, these characteristics define a lake's trophic state and have been combined into a trophic state indicator (TSI) that can be aggregated nationally by computing a frequency distribution of trophic states across lakes. The frequency distribution of trophic states (but not an average of TSI values) should be used as a national-level indicator, because changes in this distribution provide the most useful information. The national indicator should also record the number of lakes that are hyper-eutrophic, in addition to reporting changes in frequency distributions of lakes and trophic states.

In addition to the above five indicators that are directly related to productivity, **soil condition** and **land use** also are related to ecosystem functioning.

Agricultural ecosystems are a large and important fraction of the land surface of the United States (and other countries). As concern about the environmental impacts of agricultural nutrient losses has grown, so has research on how to manage nutrients to avoid or reduce those losses while maintaining productivity and profitability. Accordingly, it is useful to have indicators of both the overall efficiency of nutrient use in the production of crops and animal products and the overall nutrient balance.

The committee recommends indicators of **nutrient-use efficiency** for both nitrogen (N) and phosphorus (P). For croplands, the indicators compare the amount of N and P removed in crop biomass per year with the amount of chemical and animal (manure) fertilizer applied plus the amount of N fixed by legumes. Because livestock production is generally less efficient in nutrient use than crop production, the committee also recommends indicators of nutrient-use efficiency for agriculture overall, which compares the N (or P) content of crops for human consumption plus the N (or P) content of animal products produced with the chemical N (or P) fertilizer applied to croplands. A high value of the indicators—near 1—indicates high efficiency, while a low number—near 0—indicates low efficiency.

The committee recommends that the same information be used to compute indicators of overall **nutrient balance**. The approach is a mass-balance one. For N, the indicator would be N fertilizer applied plus N in animal waste (manure) applied plus N fixed by legumes minus N removed in harvested crops. The national-level indicators would integrate nutrient-use efficiency and the contributions from all natural processes. Thus, the indicator of nutrient-use efficiency would allow monitoring of agricultural practices and would recognize improvements in those practices; the overall nutrient balance would measure the overall environmental loadings of nutrients.

TIMING AND COST OF IMPLEMENTING THE COMMITTEE'S RECOMMENDATIONS

Much of the information that is required as input for the above indicators is already being collected at regional scales, and in some cases even at national scales. Nonetheless, full development and implementation of the indicators will be expensive and will take some time, especially for the recommended indicators of species diversity. For this reason, the committee recommends a sequential approach to the development and implementation of the indicators. Because the national land-cover indicator has general importance and because it is essential input for some of the other indicators, the land-cover indicator should be implemented first.

LOCAL AND REGIONAL INDICATORS

Indicators are needed to inform us about ecological status and trends at all spatial and temporal scales. Indeed, many indicators are useful at several scales. In addition, most policy and management decisions are made at scales defined by laws and regulations established by political entities, such as local municipalities, counties, states, and the federal government. Although the committee focused its attention on the national-level ecological indicators recommended in Chapter 4, the methods used to select and formulate those indicators are equally applicable to indicators designed for use at smaller spatial scales. Further, many national-level indicators can be reported at various levels of disaggregation to serve as regional ecological indicators. In Chapter 5, we examine a number of local and regional indicators that we judge to be especially important, and show how they can be computed and interpreted.

Productivity Indicators

In addition to a national-level indicator of ecosystem productivity, it is also useful to have indicators specifically designed to capture the performance of particular ecosystem types. In Chapter 5, we give examples of indicators for forested ecosystems; these are described below. Similar indicators can and should be developed for other ecosystem types, such as grasslands, savannas, deserts, and wetlands.

For regional forest indicators, we recommend indicators of productivity and species diversity, structural diversity, and sustainability. These attributes support the continued provision of the following goods and services from forests: wood and wood products, opportunities for recreation, tourism, and aesthetic enjoyment, maintenance of wildlife resources, control of erosion and nutrient losses to surface waters, and mitigation of greenhouse-gas emissions. The most valuable indicators for forests are those that can provide early warning of adverse trends in productivity, species diversity, and structural diversity.

We recommend that the following forest indicators be given high priority: (1) productivity and tree species diversity, (2) soils, (3) light penetration, (4) foliage-height profiles, (5) crown condition, and (6) physical damage to trees. These indicators can be assessed using data that can be collected easily in the field. In addition, the data can be used to calculate other synthetic indices (such as various diversity indices) later in the laboratory or office. They can easily be incorporated into existing inventory programs.

Indicators of Species Diversity

In addition to the national indicators of the status of species diversity recommended in Chapter 4, the nation needs indicators to evaluate the diversity status of a local area, such as a national park or an area exploited for human use. For evaluating the diversity status of such areas, we recommend three indicators: independence of the area, species density, and deficiency of natural diversity.

Although we tried to reduce the number of these indicators of diversity, and have grounded them in a single well-researched power law, all three are needed because they each inform us about different aspects of diversity.

An Indicator of Independence

This indicator assesses the degree to which the species richness of an area depends on immigration of individuals from surrounding areas. Two types of species contribute to local diversity. The first consists of *source* species, whose births exceed their deaths in the area and thus they can provide individuals to populate surrounding areas. The other type, *sink* species, are present only because immigrants compensate for their excess of deaths over births in the area. Isolating an area reduces immigration and therefore sink species will eventually disappear from the isolate. The indicator of the independence of a local area is computed using the expected values of species diversity for a large area of the same vegetation type as the area under consideration. It provides an estimate of the number of sink species in the sample but it does not specify them. These species need to be identified with traditional demographic techniques.

An Indicator of Species Density

This indicator assesses whether an area supports more or fewer species than a reasonably defined reference area does. Managers typically wish to optimize the value of their reserves. It might appear that the more species housed in a reserve, the better its condition, but this is not necessarily true. The reason is that changing patterns of land use can squeeze more species into a smaller area that cannot support so many species. As result, species will be lost from the area. The indicator signals whether diversity in the area is likely to increase, decrease, or remain the same, and it estimates the probable final diversity of the area.

Indicators of Deficiency in Natural Diversity

This indicator assesses the degree to which a site preserves exotic species of little or no conservation value rather than valued native species. When human uses dominate a landscape, natural assemblages of species disappear, but they are in part replaced by exotic species. In Chapter 4, we recommended a national indicator of native species diversity, to indicate the degree to which exotics have replaced native species. A local indicator that quantifies this tendency is also needed.

Three factors contribute to the extraordinary abundance of a few species in anthropogenic environments:

- Exotics may have had more time to adjust to us.
- Exotics may have escaped many of their natural predators.
- Only a subset of native species (the tolerants) are preadapted to "degraded" environments.

To evaluate the deficiency of diversity in an area, the raw value of species density in the area is decreased by subtracting exotic species that follow human settlement and tolerant natives that would thrive anywhere. The remaining native species density provides an estimate of the value of a site in supporting biological diversity.

The Index of Biotic Integrity: An Indicator of Species Diversity of Aquatic Ecosystems

Additive multimetric indicators have been developed and used to compare the species diversity of aquatic systems with what would be in those systems in the absence of human-caused perturbations (sometimes called *appropriate* diversity). The most widely used multimetric indicator is the Index of Biotic Integrity (IBI). The use of IBI requires general agreement about which organisms indicate poor or good ecological and water characteristics by their abundance or absence. The IBI provides a method for quantifying those qualitative assessments. The IBI is primarily a community-level rather than an ecosystem indicator because it is based on taxonomic assemblages within specific phylogenetic groups and specific biogeographic regions. The original IBI was developed for freshwater fish communities in streams in the Midwest. Recently, similar indicators have been applied to freshwater benthic macroinvertebrate communities in several regions (and even to some terrestrial communities).

CARE AND HANDLING OF ENVIRONMENTAL DATA

To ensure the accuracy and credibility of environmental indicators, procedures need to be established and maintained to monitor input data; to standardize measurements; to cross-calibrate instruments, especially when measurement technology is changing; and to document methods so that people not associated with the original data collection can reproduce the methods. Data used as input for national-level indicators should be archived in a highly disaggregated form so that these data are available for computing a variety of regional and local indicators.

RESEARCH

Although the indicators recommended by the committee are well grounded in theory and supported by extensive data, further research and development are needed to enhance the precision and interpretation of these indicators, as well as the identification and development of new indicators. Research might also suggest new indicators that are better than or that can be added to the set of indicators then in use. Research is especially needed on unusually sensitive species and processes, microbial communities, keystone species, and the temporal and spatial behavior of indicators and how their variability is influenced by underlying ecological interactions. For some recommended indicators, especially the ones that measure land cover, land use, and species diversity, further work is needed on how best to operationalize the indicators and to help identify future research plans. This work, which should include one or more workshops, should include academic scientists, practitioners, agency scientists, and other interested parties.

1

Introduction

Indicators are designed to provide clear signals about something of interest. Indicators communicate information about the status of things, and, when recorded over time, can yield valuable information about changes or trends. The bar on a thermometer indicates the temperature of a room; the light on an appliance indicates that it is turned on; the gauge on a gasoline tank indicates the amount of remaining fuel; blooms of cyanobacteria of the genus *Oscillatoria* in temperate-zone lakes indicate that serious pollution problems are developing. The values of an indicator over time can inform decisions about whether an intervention is desirable or necessary, which of various interventions might yield the best results, and how successful interventions have been. Indicators therefore can and should guide policy and help direct research.

WHY ARE ECOLOGICAL INDICATORS NEEDED?

Not everything in the environment can be measured; indicators are needed because "... of a very practical problem: too many needs, too few funds" (Jarvinen 1985). Good indicators can reveal the more significant ecological changes with the most efficient use of resources. During the coming decades, the nation will need to develop and implement environmental policies that are likely to have major societal and economic effects. If these policy decisions are to be wise and achieve their desired results, decision makers will need data that capture the most critical dynamics of ecological systems and the changes in their functioning. Good indicators serve that purpose.

Indicators usually serve as clues that something more fundamental or complicated is happening than what is actually measured. Abnormal blood pressure signals that some physiological process is not functioning properly, although it may not indicate which process is malfunctioning or why. Recording indicators over time may also signal changes or trends that are difficult to detect immediately, as when cirrus clouds, or a falling barometer, indicate the approach of a storm.

Experience amply demonstrates the power of indicators to influence human behavior. Changes in economic indicators appear to be especially motivating. When the Dow Jones index rises or falls, thousands of citizens make new financial decisions or reconsider old ones. Thousands of people pay close attention to changes in the gross domestic product. People respond to these highly aggregated indicators because they believe that these signs reveal something important about current conditions and are useful predictors of future trends and conditions. People also believe they understand what these indicators mean and what they predict.

Currently, no indicators of ecological status or trends have the stature of the most influential economic indicators, although physical indicators, such as global mean temperature, sea surface temperatures, and atmospheric carbon dioxide concentrations, are attracting considerable attention. Developing indicators of comparable power for ecological processes will help focus attention on environmental conditions, attention that may in turn stimulate significant and informed political action. Such ecological indicators are needed as yardsticks to measure public policies and their performance. These indicators must provide information in a simpler, more comprehensible form than the complex statistics usually assembled on ecological issues, and the relationship between these indicators and the complex phenomena they represent must be evident.

During recent decades, because of growing environmental concerns, increasing efforts have been devoted to developing reliable and comprehensive environmental indicators. Many countries and U.S. states now publish annual "State of the Environment" reports. International organizations, such as the United Nations Environment Programme (UNEP), analyze and report various types of environmental information. Because these reports contain large amounts of information that is difficult to digest, they often have not had much influence on decision makers. To simplify such information and make it more useful to decision makers, the Canadian government began to develop environmental indicator concepts in the late 1980s (Environment Canada 1991). The Dutch government initiated similar efforts in 1987 (Government of the Netherlands 1991). Agenda 21 of the United Nations Conference on Environment and

Development in Rio de Janeiro, Brazil, called for the development of environmental indicators to ascertain whether developmental trajectories were in fact sustainable.

In 1993, the United Nations Statistical Division (UNSTAT) and UNEP formed a Consultative Expert Group Meeting on Environment and Sustainable Development Indicators in Geneva to survey the variety of approaches being used to develop indicators. In 1994, the World Bank convened a workshop to find common ground in formulating indicators of sustainable development; and in 1995 an international policy conference was hosted by the Belgian and Costa Rican governments, together with UNEP and the Scientific Committee on Problems of the Environment (SCOPE), to seek consensus on international uses of environmental indicators (see Hammond et al. [1995] for a historical summary and references). The Santiago Declaration of February 3, 1995 (reprinted in the *Journal of Forestry*, April 1995, pp. 18-21) included criteria and indicators for the conservation and sustainable management of temperate and boreal forests. Thirteen countries that comprise the bulk of the world's temperate and boreal forests, including the United States, have agreed to monitor and report on these indicators.

Over these same years, United States federal agencies, such as the Forest Service and other agencies within the Department of Agriculture, the Fish and Wildlife Service, the National Oceanic and Atmospheric Administration, and the U.S. Environmental Protection Agency (EPA) also engaged in efforts to develop environmental indicators. Private organizations developed indicators that helped launch "green" national accounting and natural resource accounting (Repetto et al. 1989, Lutz and El-Serafy 1989), and various projects were aimed at certifying that forest products were harvested in a sustainable manner (Forest Stewardship Council 1996).

These projects have all laid important foundations for developing comprehensive, national-level indicators to inform major public policy decisions. The committee's work builds on this important work. However, the committee's focus is different from those of most previous studies because it concentrated exclusively on ecological indicators and sought to identify key ecological processes and patterns for which data could be collected as a basis for national indicators. This focus led the committee to identify the smallest number of indicators that could capture key trends in the nation's ecosystems. The committee did not propose indicators of stressors, preferring to focus on indicators of ecological conditions, most of which are influenced by multiple stressors. Some of the indicators the committee recommends, such as productivity, nutrient-use efficiency, and soil organic matter, have been recommended by others and are to some degree already in use. Others, especially those dealing with biological diversity, have not been proposed in the form proposed here. The relationship of the proposed indicators to previous recommendations is discussed in the rationale for each indicator (Chapter 4).

THIS STUDY

The Committee to Evaluate Indicators for Monitoring Aquatic and Terrestrial Environments (Indicators Committee) was formed in response to a request from the EPA for a critical scientific evaluation of indicators to monitor environmental change. Part of the motivation for the request was an earlier National Research Council (NRC) review of the EPA's Environmental Monitoring and Assessment Program (EMAP). In its overall evaluation of that program, the NRC (1995a) recommended that EMAP develop a focused indicator-research program, that each EMAP resource group develop a mechanistic conceptual model to underlie its indicators, that EMAP provide program-wide guidance for indicator development to ensure at least a consistent philosophy behind the indicators developed for the various resource groups, and that EMAP evaluate each indicator at incrementally larger spatial scales. Recognizing the difficulty of implementing these recommendations, then Assistant Administrator of the EPA's Office of Research and Development Robert Huggett requested the NRC's help. In his request, Dr. Huggett also emphasized the broad need for useful indicators throughout federal agencies and even beyond them. Therefore, while the committee hopes this report is useful to EMAP, its intended audience is much broader.

Indicators of physical aspects of our environment have received considerable attention and many of them are now in wide use in the United States and elsewhere. However, ecological indicators do not command such national confidence and use, although many have been developed and used regionally and locally. Therefore, the committee devoted most of its attention to ecological measurements that can be aggregated into useful nationwide indicators. The process and criteria that the committee used to choose these indicators should be applicable at many spatial scales, and in Chapter 5 we suggest procedures for developing regional indicators, with some examples.

The task that faced the committee was enormous, and its financial and human resources were finite. To keep its task manageable, the committee focused mainly on terrestrial, freshwater, and estuarine, but not marine, environments. Additionally, it is not obvious how data on marine environments might be aggregated into national indicators. The need for marine indicators is great, and the committee hopes that different people and insights will be brought to bear on this important task. We believe that the methods and criteria presented in this report will be useful to the development of marine ecological indicators at a national scale as well as regional and local indicators in all environments. The committee was not asked to and did not review any existing indicator-development programs, including the EPA's extensive program. It did benefit from much previous work, including work performed or supported by the EPA.

Indicators differ in their generality and their spatial and temporal scales. Indicators of the status of a particular lake may be highly relevant to local decisions about regulating pollutant discharges into the lake and its upstream watershed, but that information may not be of much use to managers living in different watersheds. Indicators are useful at many levels—community, state, regional, national, and international—and new indicators need to be developed at all such scales.

Ecological indicators, the charge to the committee and focus of this report, are urgently needed. Little attention is given here to many other important types of environmental indicators, such as those relating to climate change, ozone depletion, acidification of precipitation, and air quality. Our selective focus does not imply that these other environmental indicators are less important.

The goals of this report are (1) to summarize sources of data that can be used to design and compute indicators, (2) to suggest criteria that should be used in selecting comprehensive indicators, (3) to provide methods for integrating complex ecological information into indicators that summarize briefly but powerfully important ecological states and changes, (4) to propose indicators that meet these criteria, and (5) to offer guidance for gathering, storing, interpreting, and communicating information for ecological monitoring.

KEY ECOLOGICAL PROCESSES AND PRODUCTS THAT PEOPLE VALUE

Typically, ecological indicators are designed to assess processes and products that have value to people. People value ecological goods and services for a variety of reasons. In this report, ecological goods and services refer not only to food, fiber, building materials, and medicines, but also to the roles of ecosystem processes in protecting watersheds, reducing the frequency and severity of floods, purifying waters, and shaping local, regional, and global climates. The term also includes the services that natural environments provide for recreation, aesthetic enjoyment, and spiritual experience.

Generally, relatively constant and predictable flows of goods and services are regarded as desirable, and so indicators are needed to track changes in the processes that maintain these flows. Because environmental outputs may depend on the contributions of different sets of organisms in different ecosystems, the organisms that maintain any given ecological process are often varied both spatially and temporally.

Another guideline for identifying useful indicators is to focus on changes in states or processes that are irreversible or that can be reversed only at extremely high costs or very slowly. An example is the extinction of species, which is clearly irreversible. Although history demonstrates that the diversity of life recovers following mass extinctions, recovery may take from one to 10 million years and lost species are not necessarily replaced by anything similar. Other ecological changes that can be reversed only very slowly are loss of soils through erosion and salination. Thus, the committee devoted considerable effort to developing indicators of biological diversity and soil status.

Changes that are difficult to reverse are sometimes associated with systems that have alternative stable states. The system tends to return to its former state following a perturbation, but if pushed too far, the system may shift to an alternative stable state. Such shifts are not necessarily undesirable, but some have been. Returning the system to its earlier condition may be extremely difficult. The best known examples involve commercial fisheries. Some fisheries have collapsed to the point at which the exploited species is rare, and from which recovery has not taken place despite the cessation of commercial fishing. Examples include changes in intertidal and subtidal ecosystems from exploitation that led to local extinction of sea otters (Simenstad et al. 1978); and the disappearance of sardines off California's coast (Lluch-Belda et al. 1989). In other cases, there have been fairly persistent changes in community or ecosystem structures, as has occurred in the Bering Sea ecosystem (NRC 1996a) and for groundfish off the New England coast (NMFS 1996, NRC 1999a). In some or all of the above cases, environmental fluctuations have been a contributing factor to the problem in addition to human exploitation.

ESTABLISHING BASELINES TO EVALUATE TRENDS

To evaluate and use indicators, it is often highly informative to compare status and trends measured by the indicator against some "reference state." Without such a baseline, it is hard to assess the magnitude of change objectively, whether the magnitude of change is important, or if any efforts at amelioration are succeeding. A reference state is an operational concept that may be based on some knowledge of the characteristics of the "natural" state. This baseline could be a "magnitude of control or reference ('baseline') variability in what is being tested" (NRC 1992). Alternatively, the reference state may be based on a maintenance of ecological processes.

The phrase *ecosystem health* has often been used to identify and characterize the reference states used to gauge environmental status and trends. A major problem with this notion is the lack of clearly defined measures by which the "health" of an ecosystem can be judged. A wheat field differs dramatically from the prairie that it replaced, but a questionable amount of insight is gained by declaring the wheat field to be "unhealthy." More recently, the concept of *biological integrity* has emerged as a replacement for the term ecosystem health because this new term avoids some of the connotations of the word *health.* The concept of biological integrity is based more on the maintenance of ecosystem processes than on identification of any specific "natural" state. The term was first used in 1972 to define the goals of the U.S. Clean Water Act, "to restore and maintain the chemical, physical, and biological integrity of the Nation's waters." Biological integrity has been defined as "the capacity to support and maintain a balanced, integrated, adaptive biological system *having the full range of elements* (genes, species, and assemblages) and processes (mutation, demography, biotic interactions, nutrient and energy dynamics, and metapopulation processes) *expected in the natural habitat of a region*" (Karr 1996, emphasis added). In other words, a set of natural processes is identified and used as a reference state for evaluating current conditions and trends.

Although reference states generally are needed to evaluate current conditions, identifying the appropriate and specific reference states is difficult. First, many ecosystems and habitats are so poorly understood that specifying "natural" states and processes is possible only within broad limits. Considerable research may be necessary before specific reference states can be developed for these systems.

Second, large-scale fluctuations characterize most ecosystems. The abundances of species may change dramatically seasonally and from year to year. Rates of photosynthesis may vary greatly in response to changes in temperature, precipitation, and soil fertility. Therefore, baselines must specify typical patterns of variation and incorporate ways of deciding whether a particular fluctuation or trend falls outside the bounds of "normal" variation. The greater the normal variability in an ecosystem, the more difficult it is to identify abnormal variation.

Third, directional change occurs for various reasons, making the choice of appropriate reference states additionally difficult. Species evolve, and their larger biological communities and ecosystems can change as a result. In addition, environmental processes such as glaciation, other climatic changes, and geological forces alter the distributions and abundances of species and biological communities. Some of these changes are so slow that they are unlikely to affect the choice of reference states or reading of indicators over periods of several hundred years, but some can occur at rates roughly similar to changes caused by human population growth and technology. The choice of reference states should be made carefully by considering the implications of adopting fixed or moving baselines. Climatological "normals," for example, are 30-year averages, usually updated every 10 years, because weather patterns change over decadal and longer scales.

For some purposes, a shifting baseline is appropriate, but a particular choice of averages might not be appropriate in the face of long-term directional trends. Using shifting baselines for environmental conditions may well also lead to a relaxation of standards: gradual environmental deterioration can pass unnoticed under such a regime, in what has been called the "shifting baseline syndrome" (Pauly 1995). For example, Trautman (1981) described Ohio in the late 18th century as characterized by a "profusion of 'durable springs and small brooks,' both flowing throughout the year, and the great amount of bog, prairie, and swamp and forest lands which were covered with water during all or much of the year." The surface water was pure and clear. The change in the distribution, availability, and quality of water in Ohio—and many other places throughout the world—since 1800 is dramatic and is an example of the shifting baseline syndrome. If the change had occurred in the past 20 years, it would have caused widespread dismay and concern, but because the change took 200 years (or before current environmental awareness), the baseline for comparison has shifted and the change seems acceptable.

While recognizing all these difficulties, the committee attempted to identify products and processes for which "normal" or "baseline" conditions may be specified with sufficient rigor that they can be used as standards to evaluate important environmental changes.

EVALUATING INDICATORS

If the nation is to adopt an ecological indicator and invest the human and financial resources necessary to gather, assemble, and interpret the data the indicator is based on, the indicator's value and rationale must be clear. Therefore, for each recommended indicator, we discuss the following subjects insofar as the available information permits:

- Why the indicator is useful.
- The ecological model that underlies the indicator.
- The range of values the indicator can take and what the values mean.
- The temporal and spatial scales over which the indicator is likely to change.
- Whether the needed input data are already being gathered, and, if so, by whom.

- If the needed data are not being gathered, what new data are needed and who should collect them.
- The probable effects of new technologies on our ability to make the required measurements and how soon significant technological changes are likely.

For each indicator, it is of course essential to periodically evaluate its usefulness, reliability, and cost-effectiveness.

REALISTIC EXPECTATIONS ABOUT THE VALUE OF INDICATORS

The value of ecological indicators rests on the premise that better understanding of what is happening in the nation's ecosystems leads to better and more effective policies for encouraging desirable changes, discouraging undesirable changes, and maintaining variability within "tolerable" limits. Indicators must be developed with the knowledge that there is much that is not known about ecosystems and how they function (see Landres {1992} for a thoughtful discussion of the value and limits of ecological indicators). For example, the number of species on Earth is not known within an order of magnitude; so the current extinction rate as a fraction of all species cannot be known, even if the number of species becoming extinct each year were known (this number is known for a few groups). Only rough estimates of the value to society of ecological goods and services are available.

With good investment of financial and human resources, some current unknowns will become better understood, but some things that would be useful to know are intrinsically unknowable. No amount of effort will provide an accurate census of the number of species on Earth or their extinction rates. The exact weather several months in advance is unknowable, as is the exact rate of change in global mean temperature as a result of greenhouse gas forcing. Future research can reduce the degree of uncertainty in these estimates, but no amount of research can eliminate uncertainty completely. Ecological indicators must be developed and used with the knowledge that substantial uncertainty will always exist.

The allocation of human and financial resources to obtain the data needed for ecological indicators should be evaluated in terms of the degree of uncertainty removed per unit of effort and the resulting marginal increase in value of the indicators. Research to reduce uncertainties that have little effect on the value of currently used indicators should clearly be given lower priority than research that reduces significant uncertainties in indicators. The committee's recommendations for research have been developed with these criteria in mind.

2

The Empirical and Conceptual Foundations of Indicators

Indicators describe and summarize: they can be used for diagnosis and warning, and they can be used to monitor change. No indicator needs to do all these things, but if one wants to know whether an indicator is of value, its intended use must be clear.

Some indicators are more oriented to describing the state of a system, others to predicting its future state (NRC 1995a). Both description and prediction have their uses. However, it is impossible to imagine a successful set of indicators that fails to describe current conditions or fails to facilitate prediction. We need to know both where we are and where we are going.

Good indicators have three key features. First, they quantify information so that its significance is more apparent. Second, they simplify information about complex phenomena to improve communication (Hammond et al. 1995). Finally, indicators are used based on the assumption that doing so is a cost-effective and accurate alternative to monitoring many individual processes, species, and so on (Landres 1992). The most difficult conceptual problem in developing indicators is to ensure that they are complete enough to capture the dynamics of key processes without being so complex that their meaning—what they indicate—is unclear. Not all indicators need to have immediate policy implications, but if they are to be policy-relevant, the relationships between them and the issues relevant to public policy choices should be clear. In addition, so many ecological indicators have been proposed and used that the costs of monitoring all of these indicators would be prohibitive.

Most input data to indicators are generated by monitoring environmental states over time, either retrospectively or prospectively. The distinction between retrospective and prospective monitoring has been described by the NRC in its review of the EPA's Environmental Monitoring and Assessment Program (EMAP) (NRC 1995a):

> Retrospective or effects-oriented monitoring is monitoring that seeks to find effects by detecting changes in status or condition of some organism, population, or community. Examples include monitoring the body temperature of a person, monitoring the productivity of a lake, monitoring the condition of foliage in forests, and so on. It is retrospective in that it is based on detecting an effect after it has occurred. It does not assume any knowledge of cause-effect relationships, although the intention is usually to try to establish a cause if an effect is found. It is EMAP's general approach.
>
> Predictive or stress-oriented monitoring is monitoring that seeks to detect the known or suspected cause of an undesirable effect (a stressor) before the effect has had a chance to occur or to become serious. Examples include monitoring the cholesterol level in a person's blood, monitoring the stress level along a geological fault, monitoring animal tissues for the presence of known carcinogens or other disease-causing compounds, and monitoring with a canary the toxic gas level in a mine. It is predictive in that the cause-effect relationship is known, so that if the cause can be detected early, the effect can be predicted before it occurs.

Both retrospective and prospective monitoring are of value in developing ecological indicators. An indicator based on retrospective monitoring, which describes the state of an ecosystem, could be useful in assessing the need for environmental management or the effectiveness of environmental programs. For example, an indicator of the condition of populations of anadromous fishes, such as salmon and shad, is the degree to which a river is regulated by dams and the kinds of dams that are present. Such an indicator, if developed quantitatively, could be used to trigger action to prevent deterioration, assess future prospects, and suggest appropriate mitigation.

Prospective monitoring is common and often required by laws. For example, monitoring concentrations of various gases in the atmosphere and contaminants in water is required by the Clean Air Act and the Clean Water Act. The nation monitors for the presence of microorganisms in drinking water, community composition in some fresh waters (IBI), and the presence of various introduced and native pests.

Identifying useful indicators of stressors rather than the stressors themselves is straightforward if the likely stressor and its probable effect are known. Monitoring birds' eggs for DDT by looking at the thickness of their shells (Buckley 1986) is a good example. The stressor—DDT and its metabolite DDE—was known; at least one of its effects—interference with avian reproduction—was known; and a good indicator—eggshell thickness—was relevant and useful. Such indicators must be developed on a case-by-case basis. No systematic way exists to predict the existence and effects of stressors before they become a problem.

In many cases, however, an adverse effect is identified whose cause is not known (e.g., recent observations of deformed frogs), or a variety of adverse effects are known to be caused by a mixture of factors whose individual roles are unknown (e.g., declining salmon populations in the Pacific Northwest, various ecosystem changes in Chesapeake Bay). In these cases, stressor-indicators cannot be developed until the stressors are known. Several possible stressors could be monitored, but there are so many that this approach is unlikely to be cost-effective or efficient. For example, although the developmental condition of frogs might well be a good indicator of some as-yet unidentified stressor, and it might be a good retrospective indicator of something important beyond the valuable information it provides about frogs themselves, the value of using deformed frogs as prospective indicators cannot be assessed without knowing why they are deformed.

Some factors are both stressors and effects. For example, soil erosion is a cause of stream sedimentation, and so erosion rate is a stressor-indicator of the condition of stream communities. But soil erosion also is an indicator of the effects of overgrazing or deforestation on terrestrial ecosystems. For these reasons, the committee focuses its recommendations on national indicators that inform about the status and trends in ecosystem extent, condition, and functioning rather than focusing specifically on indicators of the stressors themselves.

Useful ecological indicators are based on clear conceptual models of the structure and functioning of the ecosystems to which they apply. The models can be empirical or theoretical, quantitative or qualitative, but, as the following discussion emphasizes, some type of model is essential to ground and rationalize all indicators.

SCIENTIFIC UNDERPINNINGS OF INDICATORS

All indicators are grounded in substantive knowledge and some sort of scientific logic. Consideration of the major varieties of such logical arguments helps assess the degree to which any given indicator is reliable.

Natural History

All indicators begin with data taken from the world. Whether these data concern the nesting season of ravens or the ozone concentration of the upper atmosphere, they are needed because it is impossible to understand the environment without observing it. Indicators based on natural history are descriptive. Even when they measure a rate, such as stream flow, they are not capable, by themselves, of predicting future conditions. To become predictive, natural history observations must be incorporated into some model of how the relevant environmental processes operate.

Paleobiology

By taking a long view of things one can determine whether current conditions and trends are within the range of variations that have occurred in the past or whether they are in some way anomalous. Are current changes in temperature and precipitation unusual? Is the current relationship between diversity and land area similar to that prevailing millions of years ago? Patterns that are repeated many times during geological history can be incorporated into the tools used to predict the future. The past can be a key to the future if the Earth has performed experiments that have yielded reasonably consistent results.

Experiments

Based on observations of the environment, scientists develop hypotheses about the relationships and processes that influence it. These hypotheses are tested experimentally in the field and in the laboratory and modified appropriately. Where they can be used, experiments are powerful tools in the development and testing of hypotheses.

Analytical Predictions

Mathematical models of environmental processes generate predictions that can be tested observationally and experimentally in the field. They also may generate predictions of future events that cannot be tested directly, but these predictions may suggest which events should be monitored to determine whether the predictions are borne out. Analytical models have been used to explain the existence of the ozone hole over the Southern Hemisphere and to predict future ranges of species under different scenarios of global climate change.

Computer Simulations

The properties of many complex ecological systems cannot be determined analytically, but they can be studied by using computer simulation models. These models allow investigators to manipulate key variables to assess the sensitivity of the ecosystem's behavior to changes in the values of these different inputs. Global climate models and models of the dynamics of complex ecosystems, which are examples of such models, have played important roles in assessing the likely consequences of human-induced changes in the environment.

USING MULTIPLE APPROACHES

The most important point to make about these different sources of scientific data is that no one source by itself is sufficient to guide the design and implementation of useful ecological indicators or to formulate sound environmental policies. Information from all these sources is needed, and the richer the array of information, the better. From a scientific perspective, indicators are most useful if they are reliable, that is, if the measurements on which they are based are repeatable and do not vary significantly depending on who gathers them.

Indicators increase in value with the time period over which their supporting data are gathered. Because the value of long-term data depends strongly on the consistency of the methods used to gather them, changes in measurement methods need to be implemented carefully. Technological advances regularly improve the speed, reliability, and accuracy with which data can be gathered. To fail to incorporate such advances is generally undesirable. Therefore, good indicators should be robust to changes in measurement technology, so that long-term data sets are internally consistent even if measurement methods have changed. The calibration of methods during periods of technological change is an essential component of integrating indicators across technological boundaries.

HISTORICAL AND PALEOECOLOGICAL DATA AS AIDS TO INDICATOR DEVELOPMENT

Paleoecology, taken here to encompass also paleontology, paleolimnology, paleogeochemistry, and paleoclimatology, is concerned with describing past ecological communities and their environments. Records of temporal change in the distribution of fossilized organisms, and of physical and chemical properties of the environment, can provide a very useful background to assess the influence on ecosystems of diverse natural and anthropogenic disturbances. These records are an extremely valuable complement to environmental monitoring, and can suggest whether the causes of past changes have been natural or anthropogenic (Charles et al. 1994). Such records, which can be gathered for a wide range of temporal and spatial scales, can help determine requirements for sampling frequency and duration.

In conjunction with monitoring programs, paleoecological indicators can help answer five main questions:

- *What were communities and ecosystems like before they were subjected to natural or human disturbances?* Evidence from pollen and seeds preserved in dated cores of lake sediments, along with other sources of information, helps to answer this question for regional and local vegetation (Brush 1986, Davis 1985, McAndrews and Boyko-Diakonow 1989). Diatoms and fossilized chlorophylls and carotenoids can be used to answer this question for lakes (Engstrom et al. 1985).
- *What have been the patterns of recovery from disturbances, and have the initial conditions been regained?* Foster (1995) used records of fossilized pollen to show that forests in central Massachusetts did not return to their composition before the arrival of European settlers following abandonment of agriculture. Recovery of lakes from severe anthropogenic acidification has been assessed by analysis of past diatom and chrysophyte species in sediment cores (Dixit et al. 1992).
- *What has been the nature and degree of variability in the past, especially the frequency of extreme events?* Heinselman (1996), using fire scars and tree-ring analysis, reconstructed the fire history of the Boundary Waters Wilderness in northern Minnesota. His analysis showed that between 1681 and 1894, the intervals between major fires ranged from 11 to 42 years. Clark (1988a, b) showed that fire frequency in this area was greater during the dry medieval warm period than during the subsequent Little Ice Age.
- *Were communities and ecosystems relatively stable, were they following trajectories of gradual change, or did they exhibit sudden fluctuations or transitions to another state?* The postglacial vegetation history of North America, established by numerous regional studies of pollen records (Wright 1971), offers extensive evidence concerning this question. In the midwestern United States, for instance, deciduous forest changed to prairie and back again during the period from 8,000 to 4,000 years ago. Renberg (1990) determined the pH history of a Swedish Lake, Lilla Öresjön, by studying its diatom stratigraphy. Many diatoms have rather narrow pH tolerances, and their sculptured shells (frustules) can be identified to species, making them excellent indicators of lake acidification. Renberg showed that Lilla Öresjön acidified slowly from neutrality 12,000 years ago to pH 5.2 2,300 years ago. It then became much less acid owing to land clearances by humans and the consequent leaching of bases into the lake. It re-acidified to about pH 4.5 during the present century owing to acid deposition.
- *Have anthropogenic perturbations been different, in degree or kind, from natural perturbations?* The history of Lilla Öresjön provides evidence that

recent anthropogenic acidification exceeded the natural acidification that preceded it for thousands of years. A difference in kind of change is illustrated by the observed eggshell thinning in birds caused by exposure to chlorinated hydrocarbons (Ratcliffe 1967), which appears to have no natural analog.

Environmental Problems for Which Paleoecological Data Have Been Useful

Paleoecological data have been used to study a wide range of environmental problems, including climate change, acid deposition, eutrophication, losses of biodiversity, fish declines and introductions, fire frequency, soil erosion, silting of lakes and estuaries, and pollution by heavy metals (e.g., lead and mercury) and trace organic molecules (e.g., DDT and toxaphene). Indeed, the understanding of most environmental problems can be improved by access to paleoecological information.

Paleoecological data have often revealed major unsuspected environmental changes, over time scales ranging from the mass extinctions observed in the fossil record to the sudden catastrophic loss of submerged macrophytes in Chesapeake Bay in the early 1970s (Brush and Davis 1984, Davis 1985). Paleoecological data have been especially important in altering views of the operation of the global climate system. Tree-ring records from the southwestern United States reveal numerous abrupt changes in regional precipitation between 800 and 2,000 years ago (Graumlich 1993, Hughes and Graumlich 1996). Likewise, close-interval paleoclimatic studies of the Greenland ice core have shown that mean annual temperatures there have changed by as much as 10°C in a few years (Grootes et al. 1993, Alley et al. 1993), and that sudden "jumps" in regional climate may be likely as changing conditions lead to a warmer climate globally (Broecker 1987, Overpeck 1996).

To anticipate "surprise" events, possible hazards will need to be monitored in a variety of environments. Attention will need to be paid to outliers in the data, which may reflect not errors in measurement but unusual, rare phenomena (Kates and Clark 1996). Thus, programs that monitor environments subject to a wide range of natural and anthropogenic stresses can benefit greatly from complementary studies of paleoecological indicators. Indeed, much useful information, such as long-term baseline records, can be gathered in no other way. Retrospective paleoecological indicators can be used prospectively to identify trends that may be useful in evaluating monitoring results.

SOURCES OF INFORMATION ABOUT CURRENT ECOLOGICAL PROCESSES

Regularly gathered information can provide the broad spatial coverage and substantial time series necessary to detect environmental trends. Monitoring key ecological information is important because the value of an indicator is reduced considerably if new baseline data must be gathered before it can be used. Because ecosystems are so variable in space and time, gathering enough data for national ecological indicators will be difficult. Fortunately, recent developments in remote-sensing technology offer new opportunities for measurements at extensive spatial scales. Because the committee recommends indicators that depend on remotely sensed information, we describe these developments in some detail.

Remote Sensing From Satellites

There are many reasons for using remotely sensed data as inputs to indicators. For many of the indicators that the committee recommends, especially for information on terrestrial ecosystem processes and some aspects of ecosystem status, remote sensing offers rapid and relatively accurate sampling. For a few remotely sensed measurements, there already is an approximately 25-year time series of reasonably well calibrated data of nearly global coverage. In addition, for measurements over large spatial areas, remote sensing offers the only affordable means of sampling. Operational costs for satellite systems are not necessarily higher than the personnel costs of large *in situ* monitoring efforts, but the capital costs of developing and launching satellite missions are extremely large.

The availability of time series of satellite measurements with well-established precision and accuracy is important in using remotely sensed data. Satellite-generated data are especially valuable if all instruments, both within and among satellite platforms, are fully calibrated and if the data are accessible and affordable. Fortunately, considerable attention has been paid to these issues.

The two types of satellite imagery most widely used by terrestrial ecologists are Landsat TM (Thematic Mapper) and AVHRR (Advanced Very High Resolution Radiometer). Both cover most of the Earth's surface, TM imagery at 30 m resolution and AVHRR imagery at 1 km resolution. Each records the intensity of radiation reflected from the planet in five to seven spectral bands, in the visible, near-infrared (NIR), and thermal infrared (IR) parts of the spectrum.

The Landsat data record, which starts in 1972, has continued through the use of three primary instruments. The first provided video images, some of which are still archived. The video system was accompanied by the Multi-Spectral Scanner, which very quickly became the preferred instrument for quantitative analysis. Since 1982, the primary instrument has been the Thematic Mapper, with six spectral bands in the visible and short-wave near-infrared, all of which have 30 m spatial resolution, and one thermal band with 120 m spatial resolution. A Landsat scene is approximately 185 km on a side; the combination of swath width and orbital characteristics of the satellite means that any spot on the Earth is revisited at about 16- to 18-day intervals. Six Landsat missions were launched, of which five achieved orbit. Landsat 5 is currently in the 13th year of its planned three-year lifetime.

The United States holds in its archives at the U.S. Geological Survey (USGS) Eros Data Center nearly complete global data from the Landsat system for the years 1972 to 1980, but the dates of cloud-free acquisitions vary geographically. Less comprehensive coverage is held in U.S. archives since 1980 because of the U.S. policy of commercializing the Landsat system (although the record for the United States itself remains fairly complete). However, 10 to 17 international ground stations, which were operating during the period of the Landsat missions, have maintained accessible data archives. Thus, in principle, global data sets could be derived for the 1980s and 1990s. The tape recorders on Landsat 5 failed several years ago, restricting its data downlink capabilities to direct transmission to ground stations (without relays to other communication satellites). This failure created holes in global coverage where ground stations are not operating, most notably Siberia, Alaska, and central Africa.

The AVHRR—a standard instrument on the National Oceanic and Atmospheric Administration (NOAA) polar-orbiting meteorological satellites—has several spectral bands in the visible, NIR, and IR parts of the spectrum. However, the AVHRR collects data with very different spatial characteristics from Landsat data, reflecting its origin as a meteorological instrument. Global-area coverage data, which are resampled on board the satellite before being transmitted, have a 4 km spatial resolution. Local-area coverage data, which are directly transmitted to the ground, have a 1 km spatial resolution. The combination of swath width of the sensor (about 1000 km) and orbital characteristics of the satellite platform achieve daily to twice-daily coverage. However, unlike Landsat TM, whose radiometric properties are known and monitored with a high degree of accuracy, the AVHRR has no provision for on-orbit calibration, and the satellite orbits have drifted substantially over the years. The data themselves are less than ideal because the instruments are poorly calibrated, and intercalibration of instruments on different platforms is also poor. Therefore, although roughly decade-long time series of truly global data exist, they must be extensively processed before they can be used for quantitative analyses of ecosystem properties and processes.

Ecosystem Processes. The most widely used satellite-derived indicator of processes is the normalized difference vegetation index (NDVI). NDVI is a direct measure of absorbed photosynthetically active radiation, because chlorophyll absorbs in the satellite's NIR band and reflects in the IR band, whereas rocks, stems, and soil reflect in both NIR and IR. The numerator of NDVI is IR minus NIR, which is large if abundant chlorophyll reduces the NIR. To compensate for the variable transparency of the atmosphere, this difference is divided by the sum NIR plus IR. Global NDVI measurements at a variety of resolutions are now available through the Internet (Kidwell 1997).

As a result of an extensive effort by the National Aeronautics and Space Administration (NASA) and NOAA, a decade-long record of 4 km resolution NDVI data with consistently processed and documented techniques is available through the AVHRR Pathfinder Program. Many individual research groups have collected and maintained archives of AVHRR and NDVI. For example, S. Los and colleagues have compiled a one-degree aggregation of 4 km resolution data for 1981 through 1990 (Los et al. in press). A complete global record at 1 km resolution is now available for the period 1993 to 1994, and continues to be collected and processed by NASA and the USGS. Collaborating ground receiving stations around the world have recently developed a more precise measure of absorbed photosynthetically active radiation than is possible with NDVI by using multiple scenes with the sun and sensor in different positions and inverting a radiative transfer model.

Once one knows the amount of leaf area in a location, it is a relatively easy matter to predict the region's net primary production (NPP). The simplest approach is to regress NDVI against measures of NPP. Generally this is done separately for each biome (e.g., Fung et al. 1987). More precise information can be obtained by using the vegetation index to parameterize a biogeochemical model, especially if there are data about the weather, topography, and soil type (which are widely available for the United States and to a lesser degree globally from NOAA and Department of Energy Web sites and on CD-ROM from these agencies). The best example of an ecosystem model driven by NDVI is the CASA model of Potter et al. (1993). This model translates monthly NDVI and climate data into predictions of primary production, carbon storage, net ecosystem production, and nitrogen mineralization. It contains only a single fitted parameter (one value for the entire globe). Although the model is relatively simple, it has an accuracy approaching direct-measurement accuracy of NPP and nitrogen mineralization—about 25 percent.

Several widely used land-surface models, initially developed to improve long-range weather forecasts, also rely heavily on satellite imagery. These models predict the transfer of matter, energy, and momentum between the land surface, vegetation, and the atmosphere at short (e.g., 20 minute) time scales. The models typically contain an enormous amount of detail at a fine scale, including the biochemistry of photosynthesis, the physiological control of stomates, the distribution of leaves in the canopy, and the vertical distribution of water in the soil. The best example of this type of model is the SiB model (SiB I, II, and III) of P. Sellers and colleagues (Sellers et al. 1997).

Because these models contain so much detail, they rely heavily on satellite imagery to fix the properties of vegetation at each location (vegetation type, albedo, leaf area, etc.). Even so, a relatively large number of parameters and functions are little more than educated guesses. Despite these limitations, models such as SiB II have a remarkable capacity to predict diurnal and seasonal patterns of production, respiration, and transpiration. SiB II also improves the predictions of those weather models that incorporate it.

The first international comparison of model simulations of NPP has recently been conducted under the auspices of the International Geosphere and Biosphere Program (IGBP). The study, which included models ranging from empirically fitted regressions to the most complex SiB II-type model, attempted to isolate the reasons for differences in estimates of ecosystem productivity and NPP. Results (Cramer et al. 1999, Cramer and Field 1999) indicate a fair degree of spread in the simulated NPP values, with some of the variation certainly attributable to differences in the underlying data sets used. However, such comparison studies are likely to lead to a greater understanding of the limitations of both the underlying data and the models themselves. The result should be better and more quantitative documentation of NPP patterns.

Vegetation Characterization and Classification. Classifying and mapping land cover and vegetation, the most common use of aerial photography, is served well by satellite imagery. The number of vegetation categories used is limited by the number of spectral bands, the resolution of the most commonly used imagery, and the experience of the interpreter. Although automated statistical techniques are used to cluster most data, there is no fully automated technique for land-cover classification.

Even simple classifications are useful as indicators of ecosystem status. For example, the rate of loss of closed-canopy humid tropical forest in Brazil was measured from the late 1970s to the late 1980s using Landsat data (Skole and Tucker, 1993) and independently verified by scientists at the Brazilian Space Agency, INPE. INPE now uses analogous methods to monitor deforestation increments on an annual time scale (Alves and Skole 1996). Classification schemes used for these analyses are primary forest, up to two categories of second growth forest, nonforest, urban areas, and water. Similarly, NDVI measurements have been used to characterize and monitor patterns of desertification in the Sahel of Africa on seasonal to interannual time scales (Nicholson et al. 1998). In both the Amazon and the Sahel, remotely sensed data provide empirical evidence for rates and directions of ecosystem change, observable on subcontinental scales.

More complex classification schemes are possible by combining imagery from several sensors with scenes from different seasons. Such techniques often depend on AVHRR because of its more frequent sampling, or on a combination of the coarse spatial resolution of AVHRR and the finer resolution of Landsat TM. For example, the dominant northern hardwood tree species can be identified using AVHRR imagery from winter, midsummer, and fall. This technology could be used to monitor changes in the abundances of common species.

Ecosystem and vegetation classification are important both for indicators and management. The USGS, through its Biological Resources Division, and in collaboration with other federal agencies, is creating a vegetation map for the United States at 1:100,000 scale from Landsat TM data. The USGS has nearly completed a major effort to map existing land cover for the United States at approximately 100 m resolution, also using Landsat TM data. Several national- and continental-scale data sets recently have been acquired by agencies with the explicit intent of promoting studies of land cover and land-cover change in the United States, the humid tropics, and North American boreal forests. These data sets are available to the scientific community for analysis. A complete 1 km resolution global land-cover product was released by the IGBP in mid-1997. Data are available at the Earth Resources Observation Systems (EROS) Data Center Distributed Active Archive Center (EDC DAAC) Web site, http://edcwww.cr.usgs.gov/landdaac/landdaac.html. This is the first global map of land cover with a hierarchical classification system produced by consistent, documented methods from a single data set. Such a map is both replicable and verifiable. Assessments of the classification are currently under way.

Vegetation and Landscape Characterization. Recent developments allow measurement of the biochemical composition of plant canopies and identification of species. Martin and Aber (1997), building on previous work by Wessman (1988), have been able to measure nitrogen and lignin concentrations in northern hardwood forest canopies at 10 nm wavelength intervals and a spatial resolution of tens of meters. In addition, hyperspectral data from low altitude measurements can be used to obtain some information on intraseasonal variations of algal species composition in lakes, i.e., the presence or absence of blue-green algae (cyanobacteria) can be detected based on the fact that these taxa contain phycocyanin pigments and other algae do not. Such data are potentially useful as inputs to vegetation classification, indicators of photosynthetic capacity, and ecosystem process models. Hyperspectral data could probably be used to monitor populations of species at unprecedented scales with high accuracy. Although several airborne hyperspectral instruments are available for research, through NASA and private industry, as yet there is no experience with hyperspectral data from orbit.

In addition to the promise of hyperspectral data, synthetic aperture radar (SAR) data have been available for several years from European and Japanese satellites and from an experimental SAR flown by NASA on the Space Shuttle. Radar backscatter is sensitive to surface wetness and the dielectric properties of the surface; some results suggest that indicators of soil moisture and vegetation biomass can be derived from SAR data for particular ecosystems. However, these techniques are generally system-specific, and the research community lacks the experience to apply them broadly. SARs are a valuable adjunct to optical sensors for vegetation classification because they can see through clouds.

Measurements of landscape spatial characteristics, which are easily derived from remote-sensing imagery using either optical or microwave sensors, are clearly important for indicators of biodiversity. For example, Skole and Tucker (1993), in addition to measuring rates of Amazonian deforestation, calculated an index of potential biological effects by assuming that a variety of different biological changes (e.g., diminished species richness) affect forests within 1 km of an edge. Biogeographic analyses of species richness on continental land scales is possible by combining spatial information and classification information that can be calculated from remote-sensing data.

Although most of the efforts to apply satellite imagery to monitoring of natural resources have focused on terrestrial systems, some current applications are relevant to hydrologic processes. For example, Landsat and SPOT satellites provide spectral information that has been used effectively to estimate soil moisture conditions (Lindsey et al. 1992) and evapotranspiration (Sado and Islam 1996), as well as soil type (Palacios-Orueta and Ustin 1996) and land use and cover (Vogelman et al. 1998), knowledge that forms the basis for estimating various hydrologic model parameters. Satellite data have also been used to assess snow coverage (e.g., Hall et al. 1995).

Satellite imagery has been used to monitor the status of aquatic systems, but the considerable potential of these techniques to produce broad geographic coverage of the nation's aquatic resources cost-effectively remains largely untapped. Satellite data have been used to monitor the effects of drought on the area of shallow lakes (Brown et al. 1977a), and they could be used to monitor wetland hydrology.

Most satellite applications to lake and reservoir monitoring involve use of spectral-reflectance data to estimate water clarity and related conditions. Reflectance of a water body is a function of optically active substances, primarily algae (and their pigments), dissolved humic matter, and suspended minerals, each of which have different spectral patterns. Water clarity (and thus spectral reflectance) in most lakes depends primarily on phytoplankton abundance in the water column, which depends generally on the level of nutrient enrichment. Reflectance data can be used to estimate chlorophyll concentrations in water (Mittenzwey et al. 1992) and thus the trophic status of lakes. Low water clarity in reservoirs often results from nonalgal turbidity (clay-like suspended solids) caused by soil erosion, and so different relationships must be developed between reflectance and trophic status in such systems (e.g., Gallie and Murtha 1993). Satellite data have been used to assess the trophic status and water quality of individual lakes (e.g., Sudhakar and Pal 1993, Chacon-Torres et al. 1992) and in a few cases the status of groups of lakes or all lakes in a region (Brown et al. 1977b, Lillesand et al. 1983, Lathrop 1992).

Landsat Thematic Mapper data are obtained at 16-day intervals across the United States. As many as five to six images are thus available for a given region or lake during the critical summer growth period (approximately the end of June to mid-September in northern states). Partial or complete cloud cover at the time of a Landsat overpass decreases the number of images available for processing, often to only one or two images per season. This frequency is insufficient for detailed assessment of time trends in individual lakes, but should be adequate for long-term monitoring of lake conditions on a regional basis. Other satellite platforms that will become available in the near future will provide more frequent coverage; the Moderate Resolution Imaging Spectroradiometer (MODIS) system, described below, will provide daily coverage of individual sites, if at lower spatial resolution (250 m, compared with 10 m resolution for Landsat). MODIS may be useful for intraseasonal monitoring of water quality in medium to large lakes.

Future Developments. Satellite-based observation systems will soon improve dramatically. With the successful launch of Landsat 7 in April 1999, satellite-based observation systems have improved dramatically. Landsat 7 has an Enhanced Thematic Mapper (ETM) on board. The ETM maintains the same spectral bands as previous Landsats, but it also has a 15 m black-and-white band, useful both in its own right and for sharpening the imagery from the 30 m bands. The ETM also has better calibration. Most importantly, the system design for Landsat 7, which again is under government control, ensures that the U.S. archive will again be truly global. The mission is designed to update the U.S.-held global archive, and the archives held by international ground stations, on approximately a seasonal basis. Data will be more affordable because the federal government will seek only to cover the costs of responding to data requests.

The Earth Observing System (EOS) AM platform—Terra—was launched in late 1999. Terra will operate five instruments, including MODIS, which is of particular relevance for monitoring ecosystem processes. MODIS will have 32 spectral channels, with stringently defined radiometric calibration. Most of the spectral channels will have 1 km spatial resolution, although a few will have 250 to 500 m resolution. With this system, a global data set can be acquired roughly every few days. In addition, the combination of this system's spectral and spatial characteristics will enable scientists to accurately calculate a variety of vegetation indices, land-cover products, fire detection products, and land-cover change indices.

In about two years, NASA will launch the Vegetation Canopy Lidar (VCL), an instrument capable of measuring foliage-height profiles everywhere on the Earth. The technology has already been tested on the Space Shuttle, and the imager itself is currently on a research aircraft. The satellite-mounted VCL will allow the monitoring of harvesting and regrowth, measuring above-ground biomass, and assessing structural habitat features important to animal species.

The commercial sector is increasingly active in providing remote-sensing information. Most of their proposed missions will seek very high spatial resolution data (1 to 10 m), either black-and-white or with a few spectral bands. Because these data provide little spectral information, they are of little use for assessing ecosystem processes. However, they have great utility for vegetation classification and analysis of land-cover changes. Recently declassified data provide an opportunity to analyze time series of land cover in some locations back into the 1960s.

Remote Sensing from Aircraft

Aerial photographs are available for most developed countries for most of the current century. Aerial photographs can provide all of the measurements available from satellites and at higher resolution, but usually in a form that is more difficult to digest. Photo interpreters are routinely employed by timber-producing industries and the U. S. Forest

Service to assess the sizes, densities, and species identities of trees in aerial photographs.

Ground-Based Measurements

Despite their great value, remote-sensing techniques do not eliminate the need for ground-based measurements. Such measurements record processes that are not detectable from afar and they are needed to "ground-truth" measurements from aircraft and especially from satellites. The most extensive and systematic system of ecological research sites is the National Science Foundation-funded network of 21 Long Term Ecological Research Sites (LTER). Several LTER sites have been in continuous operation for 15 to 20 years. The LTER sites span a range of ecosystem types on U.S. territory from arctic tundra at Toolik Lake in Alaska's Brooks Range to tropical rainforest at the Luquillo Experimental Forest near San Juan, Puerto Rico (see http://lternet.edu/network/sites/). Although each site has a different investigator-driven mission, several measurements are made at each site every year. The results, which include estimates of primary production, nitrogen mineralization rates, standing crop, abundances of most soil cations, detritus production, and censuses of dominant plant species, are available in standard form.

A second source of information that has been collected systematically for more than 50 years is the U. S. Forest Service's Continuous Forest Inventory and Analysis (FIA), which is used by the Forest Service to set timber-management policy. This system represents several thousand plots on which every tree greater than 10 cm in diameter is measured every ten years. These data obviously could be used to detect trends in the abundances of species and information about primary production. However, the usefulness of the data is reduced by the large number of errors in the archived data and the archaic format of the magnetic tape on which most of the data are stored. Precise knowledge of plot locations is not widely available to avoid the possibility that someone would manipulate the plots to affect future forest policy.

With a few exceptions, formal censuses of animal species are local and short-term. The amateur (National Audubon Society) Christmas Bird Count (CBC), begun in the winter of 1900-1901, and the North American Breeding Bird Survey (BBS), launched in 1966 by the Migratory Bird Population Station in Laurel, Maryland (now the USGS Patuxent Environmental Science Center), are conducted annually and cover the continental United States and parts of Canada (Root and McDaniel 1995, Peterjohn et al. 1995, Sauer et al. 1997). Data from both the CBC and BBS are now compiled and are available on the Internet at USGS and Cornell Laboratory of Ornithology Web sites (BBS at http://www.mbr-pwrc.usgs.gov/

bbs.html; CBC at http://birdsource.tc.cornell.edu/cbcdata/). Root (1993) and her coworkers have shown that these data are sufficiently reliable to detect temporal and spatial trends. In 1975, the Xerxes Society started the annual Fourth of July Butterfly Count (FJC), now administered by the North American Butterfly Association. The FJC is modeled after the CBC and provides data that, when used carefully, are valuable for the study of status and trends of rare and widely distributed species (Swengel 1995). The U.S. Department of Agriculture, in cooperation with other federal agencies, funds systematic studies of crop and timber pests through such programs as the Forest Health Monitoring Program (FHM) (USDA Forest Service 1994, see http://willow.ncfes.umn.edu/fhm/publicat.htm). Most Fish and Game departments census game fish, birds, and mammals. Finally, all officially endangered species are periodically censused and their current ranges are known by county (Dobson et al. 1997).

MODELS TO ASSESS ECOSYSTEM FUNCTIONING

For many years, ecosystems have been studied to determine patterns (community structure, biogeography) and processes (energy flow, nutrient cycling, stream flow, and oxygen content), but the measurements and accompanying models focus on relatively small scales. More recently, attention has also turned to processes operating at watershed scales that link upstream and downstream communities and terrestrial and aquatic communities.

Conceptual Models

Conceptual models have been significant in the development of ecosystem ecology, especially in limnology, where they have been used extensively for more than a century. For example, the concept of a lake as a microcosm (Forbes 1887) introduced the ecosystem approach to ecology and identified the processes that are still the primary foci of ecosystem ecology: energy flow, elemental cycling, production and decomposition of organic matter, and food web interactions. Lindeman's (1942) trophic-dynamic model of energy flow through the food web of Cedar Lake introduced to ecology such important topics as energy transfer efficiency, relationships between production and decomposition, and physical and chemical constraints on biological production.

Conceptual models help identify key links between ecosystem components and serve as the basis for developing quantitative models. Flow-diagram models are widely used to describe nutrient cycles, food webs, and energy flows in ecosystems. The river-continuum model is the most important conceptual model for river ecosystems. It is based on the fact that stream ecosystems undergo predictable physical and biological changes from headwaters to mouth (Vannote et al. 1980). In headwaters, terrestrial ecosystems are the principal source of the organic matter that provides energy for stream organisms. Further downstream, these extrinsic sources are increasingly replaced by instream primary production. Thus, headwaters are dominated by organisms capable of processing leaves and wood, whereas most of the consumers in downstream communities depend on instream photosynthetic plants.

The river-continuum model provides qualitative predictions about the kinds of species expected in a particular stream reach and region. The model links the physical structure of streams and expected biota, and provides a conceptual foundation for many biologically based stream monitoring approaches. Extensive data are available for mid-reaches of wadable streams, but few data are available for rapidly changing headwaters and large rivers (but see Patrick et al. [1967] for a valiant attempt to survey large rivers).

Other conceptual models have been developed to assist the design of biotic indicators and to evaluate the effects of oxygen-demanding wastes on stream ecosystems (Metcalfe 1989, Cairns and Pratt 1993). These wastes increase concentrations of fine organic particles and decrease oxygen concentrations. Among such indices are the Saprobien system (Kolkwitz and Marsson 1909) and models based on the tolerances of species to changes in turbidity and oxygen concentrations (Hilsenhoff 1982, 1987).

Conceptual models that focus on specific groups of organisms (e.g., fishes) have been used to determine minimum conditions for survival of recreationally or commercially important species. Models based on the instream flow increment method (IFIM) seek to determine stream flows necessary to support species of concern (Bovee 1996). IFIM models compare the proportional use of habitats by a species with the proportional availability of particular water velocities. Other habitat-suitability models have been developed by the U.S. Fish and Wildlife Service.

Recently, biogeographic models, based on expectations of regional distributions of species in intact habitats (Karr et al. 1985, see Chapter 1), have been used to assess the condition of stream biotas. The best known model of this type is the earlier-mentioned Index of Biotic Integrity (IBI, discussed further in chapters 4 and 5), which was first applied to fish communities and later extended to stream macrobenthos and diatom communities. IBI models have been developed and used in the upper Midwest (Illinois and Ohio), the South (Arkansas), and the West (Oregon). Many state agencies are using IBI analyses to develop biological criteria for evaluating the status of stream ecosystems (e.g., Whittier and Rankin 1992). IBI models and the analytical procedures that support them are robust and capable of detecting many changes in community composition, although often they cannot distinguish among causes of degradation. Individual IBI models are region-specific and depend on availability of extensive data on species distributions in the focal region.

Empirical Models

Empirical models are used to describe quantitative associations between variables or sets of variables and to predict the values of variables from measured values of other variables. These models are capable of making useful predictions even when cause-effect relationships between predictor and predicted variables are poorly understood.

Examples of useful empirically determined relationships are the positive correlation between chlorophyll *a* levels (a measure of algal biomass) and total phosphorus concentration (the nutrient assumed to be limiting algal growth) (Sakamoto 1966, Dillon and Rigler 1974a); the negative correlation between Secchi-disk transparency (a simple measure of water clarity) and chlorophyll *a* levels (Carlson 1977); the positive correlation between mercury concentrations in fish and their size (Lathrop 1992); and the correlation between log K_{ow} (the octanol-water partition coefficient) for various synthetic organic compounds and various measures of bioaccumulation and microbial degradability of such compounds (see Brezonik 1994 for an extensive review).

Semi-empirical models typically employ major simplifying assumptions to portray process mechanisms. Short-term variations and detailed spatial patterns in ecosystem conditions are not generated by the outputs of these models. Often called reactor models, they have been most extensively developed for lakes, which are treated as completely mixed tank reactors. Both inputs and losses of the substance being modeled are assumed to be constant or to be simple first-order processes. Reactor models were first applied by Vollenweider (1969, 1975) to analyze lake responses to phosphorus inputs. Coupled with empirical relationships between average concentrations of total phosphorus in the water column, summertime chlorophyll *a* levels, and Secchi-disk transparency measures, these models were successful in describing gross features of lake eutrophication. They were also used to develop phosphorus-loading criteria, at values above which the water quality of a lake would be expected to degrade (Vollenweider 1975, 1976; Dillon and Rigler 1974b, Baker et al. 1981). The model BATHTUB (Walker 1987) is a computerized version of the reactor model for lakes and reservoirs.

Reactor models have also been used to describe sulfate reduction and alkalinity generation in acid-sensitive lakes (Baker and Brezonik 1988, Kelly et al. 1988), retention of humic matter in bog lakes (Engstrom et al.

1988), and concentrations of various synthetic organic contaminants in lakes and reservoirs (Schnoor 1981).

The advantages of reactor models include their simplicity and low data requirements. They require few coefficients and small amounts of physical information on the system being modeled. Their disadvantages, which also stem from their simplicity, are that their coefficients lack simple physical meaning and must be determined empirically, that they are unable to capture short-term dynamics, and that they lack explicit ties between modeled output of substances and biotic responses.

Compartment models are similar to reactor models and use many of the same mathematical formalisms (Brezonik 1994). The compartments in such models generally represent mass quantities (or reservoirs) of substances or elements within discrete biotic and abiotic components. Flows of substances between compartments are expressed as simple first- or second-order differential equations. Such models have been used to describe in-lake dynamics of phosphorus cycling, including regeneration rates of inorganic P from organic P by zooplankton and microbial decomposers (Lyche et al. 1996). On a much broader scale, compartment models have been used to describe global cycling of carbon and phosphorus among major biotic and abiotic reservoirs (Lasaga 1985).

The forerunner of all deterministic water quality models is the Streeter-Phelps model for dissolved oxygen in rivers (Streeter and Phelps 1925). Developed in the mid-1920s, long before the advent of computers, the model expressed changes in oxygen concentration in a river as the difference between a loss term representing biochemical oxygen demand, caused by microbial degradation of organic matter, and a source term representing atmospheric re-aeration. Other source and sink terms, such as planktonic primary production and sediment oxygen demand, later were added to the model; it was computerized in the 1960s. Currently it can be applied to complicated river-estuarine systems and can produce time-varying output at any desired distance along the stream.

Simulation Models

The most advanced simulation models for water quality (e.g., HydroQual, Inc. 1991, 1998; Jin et al. 1998) are capable of modeling riverine-lake and riverine-estuarine ecosystems in three spatial dimensions at integration times on the order of minutes. They can produce output as daily averages over a year or several-year period for a wide range of physical, chemical, and biological variables, including water elevation (lakes) or flow (rivers), temperature, concentrations of inorganic nitrogen forms, dissolved organic phosphate, dissolved organic N and P, particulate N and P, silica, inorganic carbon, pH, dissolved and particulate organic carbon, and phyto- and zooplankton biomass.

Although some early simulation models included fishes, these models' outputs failed to accurately portray fish population dynamics. Also, current simulation models of aquatic ecosystems typically exclude benthic invertebrate and macrophyte populations. These components of aquatic ecosystems are difficult to model for several reasons: these populations are influenced by many factors other than food availability (the major driver in the models); the models simplify life cycles of organisms; and the rates at which the ecosystem populations fluctuate are often much slower than the rates of changes in other processes in the models.

Models that do not attempt to simulate variations in all three spatial dimensions have much simpler data requirements and are more practical for long-term assessments of biological processes. The one-dimensional lake simulation model MINLAKE (Riley and Stefan 1987) ignores areal variability, but treats important vertical variations, and is suitable for modeling relatively small lakes. It accurately simulates seasonal patterns of thermal stratification in temperate lakes and is fairly successful in simulating temporal trends in vertical profiles for a variety of chemicals, including dissolved oxygen. The model has been used to predict the effects of climate warming on cold-water and warm-water fish communities in Midwest lakes (Stefan et al. 1995, 1996).

A number of process models provide well-tested empirical estimates of changes based on point or diffuse source inputs of oxygen-demanding wastes (Beck 1987, Thomann and Mueller 1987). They require only measures of stream discharge and estimates of degradation and improvement parameters; no biological data are used.

Landscape Models

Spatial models of landscapes using geographic information systems (GIS) are used to make a variety of predictions about landscape changes and to indicate how those changes may affect stream ecosystems. For example, soil erosion from landscapes is predicted by the Universal Soil Loss Equation (USLE, Wischmeier and Smith 1978). The Area Nonpoint Source Watershed Environmental Response Simulation (ANSWERS; Beasley and Huggins 1982) simulates surface runoff and erosion in agricultural watersheds. This model incorporates data on overland flows that are not included in USLE. The Agricultural Nonpoint Source (AGNPS) model simulates runoff and sediment and nutrient transport across a range of watershed sizes (Young et al. 1989). This model uses grid cells to evaluate hydrology and material transport, and incorporates data on streambanks, eroded gullies, and nutrient sources. Flowpath models, such as TOPMODEL (Beven and Kirkby 1979, Beven 1997), make pixel-by-pixel estimates of conditions using land-surface data inputs. The Regional HydroEcological Simulation System (RHESSys; Band et al. 1991, 1993) and CENTURY model (Parton et al. 1992, 1994) are also widely used ecosystem models, the former including hydrology and the latter with a focus on soil organic matter (SOM). Despite the extensive development of these and many other ecological simulation and process models, the linkages between landscape processes and stream biota remain poorly understood. Understanding these connections is difficult because time lags between landscape changes and instream responses are highly variable.

THE COMMITTEE'S CONCEPTUAL MODEL FOR CHOOSING INDICATORS

To guide its selection of national ecological indicators, the committee assessed the current status of empirical and conceptual knowledge of the factors that most strongly influence ecosystem functioning. With a few local exceptions, terrestrial and freshwater ecosystems are open systems powered by sunlight. Solar energy is incorporated into ecosystems by photosynthesis, which is carried out by green plants, protists, and photosynthetic bacteria. The goods and services that ecosystems provide to humans depend directly or indirectly on ecosystem productivity, i.e., their ability to capture solar energy and store it as carbon-based molecules. Therefore, the committee recommends several indicators of ecosystem productivity.

The rate of capture of solar energy by photosynthesis is called primary productivity. Primary productivity is strongly influenced by temperature, moisture, soil fertility, and the structure and composition of ecological communities. Information on these factors can be used as the basis for accurate estimates of the primary productivity of most ecosystems. Therefore, useful indicators of ecological conditions and the productivity of ecosystems are based on data about these factors.

Extensive climatic data are already being gathered and are available to be incorporated into models of ecosystem performance. Rates of photosynthesis are measured locally for different types of ecosystems. To calculate the overall status and productivity of the nation's ecosystems, information is needed on the extent of each of the major types of ecosystem used to determine photosynthetic rates.

The condition for maintenance of soil fertility, a key determinant of productivity, is that inputs and losses of nutrients must be balanced. In natural ecosystems, new nutrients are made available by weathering of rocks and soils and by atmospheric deposition. In most agroecosystems, natural nutrient inputs are supplemented, sometimes massively, by application of fertilizers. Indicators of fertility can use data from direct measurements of nutrient concentrations in soils and of nutrient exports to other ecosystems.

Rates of photosynthesis are strongly influenced by the structure and composition of the species in the ecosystem. Structure is important because more complex vegetation may be able to intercept more sunlight than structurally simpler vegetation. The species composition of biological communities influences primary production in part because species differ in their abilities to photosynthesize under different weather conditions. Although relationships between ecosystem productivity and the number of species in the system are as yet poorly understood, it is clear that without some minimal number of species, ecosystems would function poorly (Grime 1997, Tilman 1996). Therefore, although relationships between species richness and ecosystem functioning cannot yet be quantified, the loss of species is a cause for concern. If one discovers that a species had great ecological or economic importance after it has disappeared, it is too late to do something about it. In addition, species are valued by societies for moral, aesthetic, and cultural reasons (Sagoff 1996), as expressed in international treaties and national laws (NRC 1999b).

Species composition also influences ecosystem performance by influencing the frequency and severity of diseases and pest outbreaks (Gunderson et al. 1995, Mooney et al. 1996). In addition, exotic species, many of which have escaped from their natural enemies, often achieve higher abundances than in their native lands and hence cause ecological problems (Drake et al. 1989). Therefore, measures of the presence of native and exotic species are important inputs to national ecological indicators.

How the committee used this conceptual model is described in Chapter 4, where we recommend a set of indicators that use data on the key factors that influence ecosystem functioning. These indicators are intended to provide the basis for a comprehensive national assessment of the current state and trends in the nation's ecosystems.

Policy Perspectives on Indicators

Indicators are most likely to be useful if they are understandable, quantifiable, and broadly applicable. They are likely to command attention if they capture changes of significance to many people in many places. Although indicators of local effects are not without value, they must be aggregated into some composite indicator if they are to serve broad policy purposes. Indicators are most policy-relevant if they are easily interpreted in terms of environmental trends or progress toward clearly articulated policy goals, and if their relevance is made clear (Landres 1992). In other words, indicators that convey information meaningful to decision makers and in a form these decision makers and the public can understand are more likely to be observed and acted on. Indicators are also more likely to be influential if they are few in number and capture key features of environmental systems in a highly condensed but understandable way. The manner in which data are aggregated to yield a small number of general indicators should be clear, especially to those who wish to understand how the indicators were developed. The reasons for choosing indicators, and the selection criteria, should also be clear (Landres 1992).

Any objective ecological indicators should be expressed numerically, so that results can be compared with those of indicators in other places and times. For the indicators to command attention and be used, the data and calculations they are based on must be credible. The choice of what indicators to use and how to define them is necessarily somewhat subjective, but the procedures for measurement and calculations associated with a particular indicator, once defined, must be clearly specified, repeatable, and as free of subjective judgments as possible. Where they are unavoidable, the sources of subjectivity should be defined and identified (Landres 1992, Susskind and Dunlap 1981). For example, the Consumer Price Index and the percent of people unemployed are calculated by well-defined rules that have been agreed on, regardless of a person's view about the value of full employment or low inflation or even the validity of these indices. Debates about these numbers do not involve who calculated them. Similarly, ecological indicators need to be based on calculations that are well defined and agreed on.

In addition to being based on credible measurements and calculations, the choice, motivation, and interpretation of indicators should be publicly trusted for them to be of greatest use. That means that the people and organizations who produce the indicators should be generally trusted (Greenwalt 1992). The committee cannot specify the best methods for achieving this goal, but notes that in at least some cases separating the responsibility for preparing indicators from responsibility for carrying out policies based on them seems to enhance trust in the indicators. For example, the Bureau of the Census has no policy-making responsibility; so, despite recent political arguments about the validity of sampling as opposed to counting everyone, the population estimates produced by the Bureau are usually trusted. Similarly, the National Weather Service has no responsibility for environmental policies, and so, beyond some scientific questions about the nature and placement of its instruments, its statistics are generally widely respected and trusted. The importance of public trust in the indicators is even more critical if ecological indicators are to be used as input for a national assessment of the state of the nation's ecosystems, as we hope they will be.

3

A Framework for Indicator Selection

Ecological indicators that describe the state of the nation's ecosystems and command broad and deep attention by the public and decision makers have not as yet been developed. The failure to achieve compelling nationwide ecological indicators, despite considerable effort, results from the complexity of ecological systems, their variability in space and time, and the great variety of human interactions with natural and modified ecosystems. But inadequate attention to the criteria that should guide development and use of all indicators has also contributed to the failure to achieve successful nationwide indicators. Many existing ecological indicators are applicable to only limited areas, to one type of ecosystem, or to the populations of one or a few species. Such indicators are likely to continue to be useful for their intended purposes, but by their very nature, they cannot serve as nationwide indicators.

Some indicators have been less useful than hoped because the measures employed are not clearly linked to underlying ecological processes. As a result, it has been difficult to interpret changes in those indicators. The input data requirements of still other indicators are so complex and extensive that the costs of gathering these data have been unsustainable or have presented barriers to the use of the indicator. For example, there is not enough taxonomic knowledge to use species richness of microorganisms in soils or algae in aquatic systems as environmental indicators, and even if there were, the cost of using them would be very high because the systems would have to be sampled so frequently.

To compute some indicators, intrinsically heterogeneous variables must be combined. Most currently used biotic indicators and indices suffer from problems of ambiguity in index scoring, combinations of unrelated measures, and severe limits on diagnostic abilities (Suter 1993). Complex issues may surround how such information should be combined, and a combined index of heterogeneous variables may be difficult to interpret (Washington 1984, Plafkin et al. 1989). These types of problems in indicator development and interpretation have plagued scientists and managers for years.

CRITERIA FOR EVALUATING INDICATORS

To avoid the pitfalls that limited the value of earlier ecological indicators and to provide a common framework for developing, describing, and evaluating indicators, the committee developed a general checklist. The checklist can be used to assess the potential importance of a proposed indicator, its properties, its domain of applicability, and its limitations, and thus how the indicator might be used. The entries in this checklist, reviewed over the remainder of this chapter, are general importance; conceptual basis; reliability; temporal and spatial scales of applicability; statistical properties; data requirements; necessary skills of collectors of the data; data quality control, archiving, and access; robustness; international compatibility; and cost-effectiveness.

General Importance

Does the indicator tell us about changes in the primary ecological and biogeochemical processes described in the committee's conceptual model? Does the indicator tell us something about major environmental changes that affect wide areas?

Not all indicators need to be of nationwide or international scope, but because the committee devoted most of its efforts to identifying, proposing, and characterizing indicators of general applicability, this criterion assumed major importance in the committee's deliberations. The domains of applicability of indicators of regional or local processes, or those that pertain to particular species of interest, also require attention. The spatial and temporal domain and the array of states and processes that an indicator reflects need to be specified and understood for all indicators.

Conceptual Basis

Is the indicator based on a well-understood and generally accepted conceptual model? Is it based on well-established scientific principles?

As we have noted, an indicator is not likely to be useful unless it is based on a conceptual model of the system to which it is applied. The conceptual model provides the rationale for the indicator, suggests how it should be computed, and enables us to understand the features of the indicator and how they change. Without a supporting model, an indicator's meaning and the right approach to interpreting it remain unclear (Landres 1992).

Reliability

Has past experience with the indicator demonstrated its reliability? What other evidence exists for its reliability?

The best evidence for the reliability of an indicator is, of course, its successful use previously. Nevertheless, all existing indicators should be analyzed retrospectively before assuming that their use should be continued. Inertia may lead to continued use of indicators that should be discarded in favor of better ones, or at least modified as a result of experience in their use and interpretation.

An indicator that is newly proposed inevitably lacks a historical record of reliability. Nonetheless, if it is based on a well-established scientific theory, and if a retrospective analysis has indicated that it probably would have informed us about important changes in an environmental process or product of concern, its reliability is provisionally established.

Still another check on the reliability of indicators is comparing them with already-used indicators that share a similar scientific rationale and employ similar input data. If those indicators have proven to be reliable, the proposed one is likely to be as well. For example, a variety of community-level diversity indicators based on stream invertebrates have been patterned after established indicators also based on the diversity of fishes in streams. Thus, the Invertebrate Community Index is essentially equivalent to the IBI (Plafkin et al. 1989).

Temporal and Spatial Scales of Applicability

Does the indicator inform us about national, regional, or local processes and products? Are the changes measured by the indicator likely to be short-term or long-term? Is the indicator sensitive enough to detect important changes but not so sensitive that signals are masked by natural variability?

To determine what an indicator indicates, the kinds of data needed to compute it, and how changes in it should be interpreted, the temporal and spatial scales of the processes measured by the indicator need to be clear (Peterson and Parker 1998). Much about an indicator's relevant spatial scale is shown by the scope of the data used to compute the indicator, but temporal scales of applicability are much more difficult to determine, although paleoecological studies may be helpful in doing so. For example, the number of taxa, the lack of good methods for delineating them, and the heterogeneity of community composition at microscopic scales make it infeasible to assess the diversity of microorganisms at present.

Over the past decade, ecologists have gained an awareness of the critical importance of scale. Virtually all metrics of ecosystems depend on spatiotemporal scale, so the explicit selection of spatial and temporal scales for indicators is essential. The field of spatial statistics and geostatistics provides an extensive foundation for the description of spatial patterns and how they change across spatial scales. Foundations for the variety of statistical methods and tests used to assess spatial descriptors are found in Bartlett (1975), Ripley (1981), Cliff and Ord (1981), and especially Cressie (1993). These methods allow identifying the scale at which a pattern exhibits the least random variation and the scale at which the pattern changes abruptly.

Spatial statistics are useful in evaluating alternative scales for indicators. All else being equal, the better scale is one at which an indicator exhibits the least stochastic variation and the weakest dependence on small changes in scale. Spatial statistical patterns might also serve as indicators of spatial qualities, such as landscape heterogeneity or habitat diversity. The primary problem with spatial statistics is their requirement for extensive input data. Except for satellite and aerial imagery, most ecological data sets lack the necessary sample sizes and spatial coverage. Without knowledge of the scaling rules, the scales at which measurements are made cannot be extrapolated to the scales at which indicators are needed. This is an issue that would benefit from a focused research program, using data from a large sampling program such as EPA's Environmental Monitoring and Assessment Program (EMAP).

Statistical Properties

In the areas of accuracy, sensitivity, precision, and robustness, has the indicator been shown to be good enough to serve its intended purpose? Can the indicator detect signals above the "noise" of normal environmental variation? Are its statistical properties understood well enough that changes in its values will have clear and unambiguous meaning?

Because ecosystems vary in space and time, indicators of their status and functioning also vary spatially and temporally. In addition, ecosystems are changing today in unprecedented ways. Lack of historical and contemporary data make it difficult to define clearly the nature and extent of these changes (NRC 1986), although paleoecological studies can be useful. Useful indicators should be able to distinguish between "normal" variation and variation that falls outside what is expected given the historical and paleoecological record.

Ecological indicators are designed so that changes in their values signal significant changes in ecosystems, changes to which attention should be paid. When a signal must be detected against a variable environment, which is true in most ecological conditions, it is essential to consider which of two types of errors is more important to avoid (Simberloff 1990). The first, known as a Type I error, is to conclude that a significant change has occurred when it has not, that is, interpreting normal environmental variability as a real change. The second, or Type II error, is to conclude that there has been no significant change when in fact such change has occurred. When a Type I error is made it may prompt an unnecessary, costly, ineffective, and possibly counterproductive response. When a Type II error is made, a needed response may not be made, which may compound a serious problem. When the needed responses are eventually initiated, they may be both more costly and less effective in alleviating the problem than if they had been undertaken earlier. Each indicator should be evaluated to assess the relative importance of Type I and Type II errors associated with its use. (Statistical issues are considered further in Appendix A.)

Data Requirements

How much and what kinds of information are necessary to permit reliable estimates of the indicator to be calculated? How many and what kinds of data are required for the indicator to detect a trend?

All indicators require input data, but what they require differs dramatically in nature and extent. Most ecological indicators depend on data gathered by means of long-term monitoring. The challenge is deciding which rates of change to watch, and to determine which of the changes observed are normal and which are not. Often changes in rates can be determined only after long periods of time. The significance of changes over short time periods is often unclear because the record is too short to characterize natural variability in the system. Temporal and spatial variation is often considered something to minimize by clever sampling design, but such variability may demonstrate the most interesting and important features of the system (Kratz et al. 1991).

Ecosystems may change very slowly, in response to factors such as changing climate, soil properties, and evolutionary changes in species; moderately slowly, as vegetation succession occurs and species ranges change; or suddenly, in response to disturbances, either natural (e.g., fire, storm, or disease) or anthropogenic (e.g., acid deposition, wetland drainage, forest clear-cutting, or species extinctions). To develop an indicator that can signal significant ecosystem changes, whether in response to natural or anthropogenic perturbations, it is essential to identify current system states—physical, chemical, and biological—in both stressed and unstressed ecosystems. Where possible, it is also desirable to establish the past states of the system through historical and paleoecological studies that provide baselines for a program of physical, chemical, and biological monitoring. Those properties of ecosystem structure and functioning that paleoecological studies have identified as significant indicators of change are prime candidates for inclusion in current monitoring programs. Information from the monitoring program can then be used to devise an indicator whose properties can be specified and interpreted.

Once an indicator is selected, monitoring must be used to increase knowledge about the likely meaning of changes in the indicator's values. Experimental studies—themselves requiring monitoring—should be used to determine whether the stress/response relationships suggested by the monitoring program are indeed causal. The use of the indicator may change as additional insights are gained into its behavior and the underlying processes that cause it to change.

Necessary Skills

What technical and conceptual skills must the collectors of data for an indicator possess? Does the collection of input data require highly technical, specialized knowledge if the data are to be accurate, or is data collection a relatively straightforward process?

An indicator capable of commanding broad attention must be based on data that are accurate and, equally important, perceived by all interested parties to be accurate. Accuracy, both real and perceived, is more likely to be achieved if it is clear how the input data are collected and that the gatherers of the data have skills appropriate to their assigned tasks. Thus, it is desirable that indicators be designed to use input data that are relatively straightforward to gather.

Because collection of data for ecological indicators (monitoring) is sometimes perceived by scientists as boring or less interesting and prestigious than "scientific research" (i.e., hypothesis-driven investigation), it is important to provide incentives for consistent and accurate data collection. Monitoring data can often have great scientific value, but not necessarily to the individuals who collect them. Much scientific research involves the painstaking and repetitive collection of data, but the potential reward of being able to answer a scientific question is often sufficient incentive to lead people to spend long hours and even sleepless nights to collect them. For this reason, as well as for the obvious scientific benefits that would accrue, the committee suggests that monitoring of ecological indicators be coupled with more focused scientific research whenever practical. This could mean taking advantage of long-term research projects (such as the National Science Foundation's Long Term Ecological Research {LTER} sites), or inviting researchers to make use of data collected by a monitoring program to test scientific hypotheses. The indicators we have proposed embody hypotheses about the functioning of ecosystems. To the degree that such hypotheses can be made explicit in the design of indicators, their development and the subsequent monitoring of them should generate a great deal of valuable scientific information. Other incentives might also help to achieve the goal of obtaining consistent and accurate monitoring data.

Robustness

For our purposes here, we define robustness in a nonstatistical sense, as an indicator's ability to yield reliable and useful numbers in the face of external perturbations. In other words, is the indicator relatively insensitive to expected sources of interference? Are technological changes likely to render the indicator irrelevant or of limited value? Can time series of measurements be continued in compatible form when measurement technologies change?

Indicators are influenced both by changes in the systems monitored and by changes in technology. During the life of an indicator, scientific advances are likely to improve understanding of the system's dynamics and hence the ability to interpret changes in the indicator. The expanded knowledge may also suggest that additional data should be included as inputs to the indicator, or that data previously collected are less relevant than they formerly appeared, and perhaps their use should be discontinued. In addition, technological advances may enable measurements to be made that are currently impossible. Other measurements may become much easier. Thus, the input data to indicators are likely to change.

Such technologically driven changes are, of course, welcome. To continue to gather data by outdated methods is undesirable. Nevertheless, because long-term data sets are essential for detecting most environmental trends, technological changes must be incorporated into monitoring programs in ways that do not destroy the continuity of the data sets or render consistent interpretation of the changes impossible. As pointed out in Chapter 2, cross-calibration of measurements is especially important for remotely sensed data.

International Compatibility

Is the indicator compatible with indicators being developed by other nations and international groups?

Not all indicators used in the United States, especially those relating to specific regions, ecosystems, or species, need to be compatible with indicators developed and used in other nations. However, national-level indicators signal changes that are likely to transcend national boundaries. Effective responses to these changes may require international action. If the signals that trigger actions are not mutually interpretable to the affected nations, appropriate responses are certain to be more difficult to mount.

Costs, Benefits, and Cost-Effectiveness

Costs and benefits associated with proposed indicators are important because resources for monitoring are limited and should be used efficiently. The costs of developing and monitoring an indicator, which can continue to accrue as the indicator is used and refined and as new data and technologies develop, can be estimated objectively. The benefits—the value of the information obtained—are more difficult to estimate. They include both its contribution to scientific progress and to improvements in societal decisions, and they also continue to accrue over time. The greater the contributions by an indicator, the higher the costs that can be justified in developing and implementing it. For example, developing and monitoring a national land-use indicator will be costly, but the committee considers that the information will be of great value and will be an essential component of other indicators.

Indicators can be judged based on several criteria, an important one of which is cost-effectiveness. If one assumes that the information an indicator yields is essential, can it be obtained for less cost in another way? If so, the indicator is not cost-effective (Landres 1992). Another criterion is more restrictive: Is the value of the information to be obtained greater than the cost of obtaining it? This is the positive-net-social-benefits test, which is difficult to apply because it requires dollar (or some proxy) measurement of the societal value of the information. The value of the information was the committee's first consideration in every indicator we recommend.

INFORMATION HANDLING AND CALIBRATION

Although tremendous strides have been made in remote sensing over the past six years, the use of this information for ecological analyses is still in its infancy. For many years, the technological challenges of handling and processing the data were so great that only the most sophisticated laboratories could use these data. Because rapid improvement, in both cost and performance, of computer hardware and software is now removing many technical impediments to the use of remotely sensed data, it is increasingly important to pay attention to the care, maintenance, and accessibility of data archives, and to the intercalibration of the remote-sensing instruments themselves.

The processes and conditions for which the committee recommends indicators operate at multiple spatial and temporal scales. Instruments that are used to measure temporal and spatial variation must be calibrated carefully to ensure confidence in ascribing changes in measurements to the ecosystems being monitored, rather than to the instruments themselves. Although simple to stipulate, achieving calibration precise enough for quantitative scientific measurements is very difficult. Delicate hardware that has been calibrated and tested on the ground, for example, is subjected to tremendous vibrational stresses during launch and then to the thermal and radiational stresses of low Earth orbit. Maintaining instrument calibration is a nontrivial task under these circumstances, but it can be achieved, as the Landsat data record shows.

For data sets that must last longer than the lifetime of any one instrument, it is equally important to ensure that successive instruments are flown for a period of overlap. As both the hardware and software of instruments evolve, one must be able to identify and quantify degradation of instrument performance and offsets due to changing satellite orbital geometry or new technology, and correct for them before further analyses are attempted.

DATA QUALITY CONTROL, ARCHIVING, AND ASSIGNMENT OF RESPONSIBILITIES

Before any indicator is adopted, substantial thought and effort need to be given to issues of data quality control, data archiving, and data access. The integrity of time series of information is vital because individual measurements acquire value only when they are compared with the same measurements from other similar ecosystems or from the same ecosystem at other times.

The data on which ecological indicators are based must be archived and available to a wide range of interested parties if the indicator is to be accepted and used to guide policy. Before any indicator is designed and used, its archival requirements must be considered carefully and accommodated.

Issues that need attention include the following:

- How and by whom will quality control over input data be ensured?
- Who are potential users of the data and how can their needs be met?
- How can the data be used to improve the models on which the indicator is based?
- How can the archival system best accommodate technological changes in both data collection and archiving methods?
- Who will coordinate and manage the archives?
- How can the system respond to complex user queries that may require new analyses and interpretation of existing data?
- How will the data storage systems be integrated with other archival systems of federal, state, and local governments?

Data Quality Control

The indicators the committee recommends are calculated from extensive underlying empirical data collected by a variety of monitoring programs. Each of these programs has some capacity to ensure that their data are of high quality, that they are archived appropriately, and that they are accessible.

No indicator of environmental quality is reliable unless the underlying data that are used to construct or calculate it are accurate. No amount of attention to data quality during the archiving and computational phases can substitute for the quality of the input data. In this critical sense, the ultimate responsibility for data quality must lie with the investigators who collect them. Therefore, a successful monitoring system must ensure that there are sufficient incentives in place for participating investigators to maintain quality checks on their data.

Clear documentation of sampling and analytical methods is necessary if future investigators are to understand exactly how each indicator was derived. Therefore, the original investigators must document their methods carefully enough so that someone not associated with the original data collection can reproduce the original sampling or analytical protocols. This requirement is particularly important as methods and instrumentation change, so that data from early parts of the time series are quantifiably comparable to data from later parts of the same time series. Individual investigators must clearly record changes in their methods, and document the influence of those changes on the measurements.

Data Archiving

A monitoring system to track ecological indicators requires archiving capabilities that provide interested parties access to the data. To design an archiving system, it must be clear whether "raw" data are to be archived, or only derived quantities such as the indicators themselves.

For indicators that are direct representations of environmental samples, the archiving job is simple: the archive simply needs to save a record of the measurements, such as phosphorus concentrations. A more difficult question is to what degree the physical samples themselves (e.g., soils, water, or plant, animal, and microbial specimens) should be archived. In general, the minimum number of physical samples saved should ensure the ability to recalibrate the entire data set, should that become necessary because of changes in sampling or analytical technologies (so-called technology drift). The costs of preserving physical samples in forms that do not decay or otherwise change must be weighed against the opportunity cost of not being able to recalibrate a data set with improved or modified measurement techniques.

Indicators of net ecosystem productivity or net primary productivity for any substantially large area are computed using some form of remotely sensed data combined with model simulations. The archive must encompass procedures to preserve not only the original data, but also the models that have been used to interpret or extrapolate from them. Governments are likely to be the source of most remotely sensed data for the next decade or so. Substantial provisions have already been made to ensure that the original digital data are archived in easily accessible and affordable forms. Little could be gained by another monitoring system to duplicate the archiving of these data. However, remote-sensing archives do not necessarily save derived products and other analyses computed from the original data, unless they are specifically funded to do so. Therefore, estimates of productivity and similar derived values should also be archived, together with pointers towards the original imagery and a complete description of the models that have been used to derive the estimates.

The complete description and availability of the models and their metadata used to derive final indicators are just as important as the availability of the underlying remote-sensing data themselves; otherwise, future comparisons might not compare the same things. The models that have been used to date to calculate such variables as ecosystem productivity and net primary productivity are active research tools, and they too will continue to evolve. Therefore, the archival task is to ensure that there is as much traceability in the models as there is in the measurements. The archive must therefore be robust enough to ensure that the time series of the indicator can be reprocessed as models improve.

Multimetric indicators face similar problems because they are calculated from sampling data. If the biological sampling data are part of ongoing research or monitoring efforts that have already made provisions for archives, there may be little need for additional archiving capabilities. However, if the biological data are not otherwise archived, then the national system must make provisions for archiving the original data, and not merely the values of the index itself.

Assignment of Responsibilities

How should the responsibilities of individual investigators be assigned and monitored? At a minimum, the monitoring systems that contribute data to indicators must include some central function that establishes and maintains appropriate standards for data quality that investigators who make the measurements must meet. The standards themselves do not need to be centrally created and unilaterally handed to investigators. Instead, the investigators themselves can discuss and develop appropriate standards that they agree to uphold. However, the central mechanisms must be adequate to ensure that when data are proposed for entry into archives, they satisfy the agreed-on standards.

This process differs from the centralized data quality control processes that have often been used in large monitoring and research programs. It envisions the central function as that of a monitor, rather than a controller. The responsibility for setting standards, for changing standards when appropriate, and for upholding standards, is the responsibility of the investigators themselves.

Implementing this function in the monitoring systems that provide data for indicators will be expensive. Experience with large field programs (e.g., the National Aeronautics and Space Administration's First ISLSCP {International Satellite Land Surface Climatology Project} Field Experiment {FIFE} , the National Aeronautics and Space Administration's Boreal Ecosystem Atmosphere Study {BOREAS} , the National Science Foundation's Long-Term Ecological Research {LTER}) suggests that the costs of the data systems that support intensive monitoring efforts can easily amount to 25 percent of the total budget. Nevertheless, the monitoring of input data is critically important and must be supported at levels that ensure that data quality and compatibility do not decline over time.

USE OF THE COMMITTEE FRAMEWORK

In the following two chapters, the committee uses the framework described in this chapter to identify a coherent set of indicators that can provide a comprehensive view of the status and trends of the nation's environment. These highly aggregated national-level indicators, which are based on well-established scientific data and models, require extensive measurements from all parts of the nation as input. As we show in Chapter 5, if the data are archived in a highly disaggregated form, they can be used for indicators designed for local and regional levels as well.

4

Indicators for National Ecological Assessments

In the preceding chapters, we have discussed the desirable characteristics of indicators, the sources of data that underlie them, the models that support them, and the criteria by which good indicators can be identified. Based on those discussions and the conceptual model we used, the committee recommends national indicators for three major categories of ecological information. These categories encompass the nation's most important ecological issues. By computing them and paying attention to them, the nation should be aware of the status of its ecosystems, be alerted to changes that may require management interventions and policy changes, and have a basis for ensuring that future generations will have access to ecosystem goods and service as rich as those enjoyed today. In some cases, noted for each indicator, some experience will need to be gained on details of the indicator's behavior, but all the indicators are based on soundly established scientific experience and principles. The proposed indicators are in general applicable to both managed (e.g., agricultural) and unmanaged ecosystems; the indicators of nutrient-use efficiency and overall nutrient balance are specific to agricultural ecosystems.

- **Information about the extent and status of the land use and cover types that together make up the nation's ecosystems.** Information about the extent of the nation's land use and cover types informs us about the extent of ecosystem types in the nation, and it is needed to calculate several other indicators. The information and technology to calculate land cover is currently available. Land use in some ways is more informative because it provides additional information on the status of areas and hence their ability to provide goods and services. Also, information on how land is used is predictive of future land cover and hence predicts the ability to provide goods and services. However, a land use indicator requires much synthesis of existing information and some new information, and thus will take longer to develop than a land cover indicator. Meanwhile, land cover can serve as a valuable indicator.

- **Information about the nation's ecological capital.** This information measures the nation's natural capital, or raw materials, both abiotic and biotic. Abiotic raw materials essential for ecosystem functioning include soil (also partly biotic) and its nutrients. Biotic raw material includes the number of species still present in the country relative to their number at the time of European contact, their distribution over today's natural and human-modified environments, and the number of species present today that are exotics, those species introduced, deliberately or inadvertently, that have become established or naturalized.

Indicators of biological capital are important for several reasons. Knowing what portion of our biological capital is native provides a sensitive measure of humans' environmental impacts, as described below. Assessing the status of the nation's biological capital is important for ethical and aesthetic reasons. Also, biological diversity is an indicator of the capacity of ecosystems to function effectively. An important but controversial theory holds that because species differ, species-rich ecosystems are more likely than species-poor ecosystems to contain some species that can thrive during an environmental perturbation (Mooney et al. 1996). As a result, species-rich systems should be buffered against disturbances and continue to perform better in fluctuating environments than species-poor systems. Empirical studies are still few, but they provide some support for the hypothesis (Tilman 1996, Tilman and Downing 1994, Naeem et al. 1994). Further investigations will clarify relationships between biological diversity and ecosystem processes. Meanwhile, it is prudent to monitor the status of the nation's biological resources.

- **Information about the functioning (performance) of the nation's ecosystems and how it is changing**. This information includes measures of productivity and other ecosystem processes. Changes in the productivity of ecosystems are generally accompanied by changes in their ability to provide goods and services. Usually, declines in productivity are undesirable, but in freshwater ecosystems, increases in productivity associated with eutrophication can be undesirable.

For each of these major categories of information we recommend indicators (Table 4.1) that are described in detail below. Although the categories apply to all ecosystems, they differ in many details for marine,

TABLE 4.1 National Indicators of Ecological Condition

Category of Ecological Information	Recommended Indicators	Reasons for Choosing Indicator
Extent and Status of the Nation's Ecosystems	Land Cover and Land Use*	Needed for calculation of most other indicators. Inform us about the overall extent of different ecosystem types.
Ecological Capital Biotic Raw Materials	Total Species Diversity	Measures nation's biological resources (what is present relative to what is expected).
	Native Species Diversity	Measures the amount of biological diversity that is native.
Ecological Capital Abiotic Raw Materials	Nutrient Runoff	Estimator of total losses of nutrients. Nutrient runoff has major effects on receiving waters.
	Soil Organic Matter*	Best single indicator of soil condition, related to erosion.
Ecological Functioning (Performance)	Productivity, including Carbon Storage, Net Primary Production (NPP), and Production Capacity	Direct measures of the amount of carbon sequestered or retained in an ecosystem (NEP), energy and carbon brought into an ecosystem (NPP), and energy-capturing capacity of ecosystems (chlorophyll).
	Lake Trophic Status	Direct measure of ability of lakes to provide goods and services.
	Stream Oxygen	Captures the balance between instream primary production and respiration.
	Soil Organic Matter*	Single most important indicator of soil quality and productivity.
	Nutrient-Use Efficiency and Nutrient Balance	Inefficient use of nutrients is costly in terms of economics and damage to ecosystems to which nutrients are discharged.
	Land Use*	Provides information about ecosystem functioning.

*Indicators in more than one category.

freshwater, and terrestrial ecosystems. Because marine ecosystems were not specifically covered by the committee's charge, we focus on terrestrial and freshwater ecosystems and the interface between them–wetlands. More research is needed for the full implementation of these indicators. We offer some suggestions for the order in which some components might best be initiated.

Although the national-level indicators we propose are highly aggregated, most of them require detailed input data. If these data are archived in a disaggregated form, the indicators can be computed in a variety of ways to provide a rich array of indicators of great local and regional value. Chapter 5 describes how that can be done and describes additional regional and local indicators.

THE EXTENT AND STATUS OF THE NATION'S ECOSYSTEMS

To estimate the capacity of U.S. ecosystems to continue to provide the goods and services that society depends on, one needs to know the status of the different types of natural and human-modified ecosystems and how much of each major type of ecosystem remains in the country. Information is also needed on what the committee calls the *matrix* that the ecosystems are in, i.e., the physical aspects of the land. Thus the indicators in this category include *land cover* and *land use; nutrient runoff* to coastal waters (a measure of loss of an element of the matrix); and *soil quality* as measured by soil organic matter.

Land Cover and Land Use

Land cover refers to the ecological status and physical structure of the vegetation on the land surface (e.g., forests, grasslands, wetlands, croplands) (Meyer and Turner 1994). However, land cover depends in part on *land use,* the way the land is used by people (e.g., a forest managed for timber, a forest used to conserve biological diversity, industrial areas, areas of human settlements) (Meyer and Turner 1994). Because changes in land use often (but not always) affect land cover, land use is itself an indicator of land cover. In general, land cover can be detected and monitored from remotely sensed imagery, but detection and classification of land use usually requires on-the-ground measurements. Often, especially in industrialized countries, information about land use can be obtained from maps and other data sources at local and regional scales, but such compilation is more labor-intensive and expensive than getting information on land cover.

Data on current land cover and trends in those values are essential for the derivation and use of most indicators the committee recommends.

Therefore, we present the land cover indicator first and recommend that top priority be given to developing this indicator as rapidly as possible. A reliable land use indicator, although more difficult to develop, is also of great importance. The mathematical tools needed to assess land cover patterns and how they change are detailed in Appendix B.

To assess this component of human impacts on the land, the committee recommends a land cover indicator to track the amount of land in each of an array of land cover types, such as croplands, forestlands, wetlands, and nature reserves.

A large fraction of the Earth's land surface is devoted to agriculture, and so agroecosystems—ecological systems that are intensively managed for the production of food or fiber—are essential components of any land cover and land use indicators. Agricultural systems are managed for high production, and typically are characterized by intensive nutrient and pesticide inputs, fast growth/harvest cycles, and low plant and animal diversity (Odum 1984, Matson et al. 1997). Negative effects that may accompany these patterns include increases in soil erosion, groundwater contamination, eutrophication of lakes and rivers, and increased resistance of pest and plant pathogens to the chemicals used to kill them (Matson et al. 1997). Nevertheless, intensively managed agroecosystems usually contain parcels of unmanaged or lightly managed areas, such as woodlots, fencerows, or riparian areas that can act both as refuges for beneficial predators of insect pests (Letourneau 1997) and as reserves for insect pests, weed seeds, plant pathogens, and alternate hosts of fungal pathogens of crops, such as cedar apple rust or crown rust of oats (Schumann 1991).

Because high-value farmland is being lost to commercial, industrial, and residential land uses (USDA 1997), the amount of agroecosystems is a component of the land cover indicator.

Both land cover and land use indicators are extremely important. Although this section focuses on land cover, because data are more readily available at present, much of it is relevant to land use, which should also soon become a usable indicator.

The land cover indicator can be applied to all environments, including those in which the "land" is submerged. Therefore, the land cover indicator recommended by the committee, includes rivers, wetlands, and riparian zones as well as dry land. The land cover indicator records the percentage of land in each of a variety of land cover categories. Every time land cover is computed, the proportional representations should be compared with those that existed at the previous recording time. To provide a useful indication of the status of the nation's lands, many categories of land cover types must be recognized and the input data entered and stored separately for each one. Because the proportion of land in each land cover category changes relatively slowly, land cover needs to be reported only once every five years, but its values need to be computed annually as inputs to other indicators.

Supporting Models and Data Requirements. The conceptual model for land cover is simple. Land cover measures the proportion of the landscape (and waterscape) occupied by each member of a set of land cover types that must, by definition, add up to the total area of the nation. The major decisions concern the number of land cover categories to recognize, how to account for their spatial configurations, and how to accommodate changes in the number and kinds of categories that are recognized. Change in the proportional representation of various land cover categories is the variable of interest. Visually representing proportional changes for a large number of land cover categories is difficult, but pie or star diagrams may serve that purpose.

More sophisticated analyses of changes in land cover categories are also possible through the use of Markovian models (Appendix B). Those models assemble the transition probabilities between categories into matrices. From the matrices the steady-state distribution of categories and the rate of approach to the steady state can be calculated. As a result, one can obtain an indication of what the landscape will eventually look like under selected policies, and how long it will take to reach that state.

Data Needs. Current capabilities of satellite imagery are sufficient for identifying a large number of categories of land cover. Very complex classification schemes are possible by combining imagery from several sensors with scenes from different seasons. As pointed out in Chapter 2, such techniques depend either on the Advanced Very High Resolution Radiometer (AVHRR), because of its more frequent sampling, or on a combination of the coarse spatial resolution of AVHRR and the finer resolution of Landsat Thermatic Mapper (TM). The forthcoming map by the U.S. Geological Survey (USGS) with a 100 meter resolution, based on Landsat TM data, also may serve as the basis for establishing categories.

The condition of the nation's flowing waters is clearly an important element of the land cover indicator. An aggregated measure of flow patterns of the nation's rivers would be useful, but the complexities of river-flow dynamics make such a measure computable and understandable only at the level of river basins. For the land cover indicator, a simpler measure—percent free-flowing (as explained below)—captures enough of the pattern to serve as a useful surrogate.

Dams, regardless of their purposes, change discharge patterns, inhibit movement of fishes, and impound sediments and their contained organic carbon, nutrients, and contaminants (Poff et al. 1997). Water is retained in reservoirs to alter the pattern of peak flows and rates of flow during storms. The timing of water release determines the timing and magnitude of seasonal runoff maxima, which often differ from those in free-flowing streams in the region. Therefore, for this reason it will be desirable for the land cover indicator to include a measure of percent free-flowing, or the length of free-flowing parts of streams and rivers divided by their total length. The percentage can be computed for each river basin and then aggregated into a single nationwide value.

Percent free-flowing can change in response to policy and management decisions. Benke (1990) estimated that of the 5,200,000 km of streams in the contiguous 48 states, only 42 streams flowed unimpeded for more than 200 km. This is much less than 1 percent of the length of the nation's rivers. Of these 42 rivers, only six, all of which are in the southeastern United States, flow to the sea. Damming of rivers may continue, at least for hydropower generation, with the greatest number of undeveloped sites in the Pacific, mountain, and northeastern states. On the other hand, sentiment is increasing for removing some dams. Data on the length of large rivers impounded behind dams can be obtained using satellite measurements, but information on dams on small rivers can be gathered only by field surveys. However, because the number of dams built or destroyed over a short period is a very small percentage of all dams, the percentage free-flowing would change extremely slowly. Updating the data base would be relatively simple once it had been compiled.

As mentioned previously, the U.S. Department of Agriculture's National Resources Inventory (NRI) provides a comprehensive assessment of the state and performance of natural and agricultural ecosystems on 800,000 sites on private lands every five years. The NRI is a comprehensive sampling of land cover, land use, soil erosion, prime farmland, wetlands, and other characteristics on nonfederal lands in the United States. These data show that although soil erosion and agricultural wetlands loss have decreased, 6 million acres of prime farmland were converted to nonagricultural uses between 1982 and 1992. NRI data can form part of a system to confirm ground truth of land use categories identified by satellite imagery.

The U.S. Geological Survey (USGS), through its Biological Resources Division, and in collaboration with other federal agencies, is creating a vegetation map for the United States at 1:100,000 scale from Landsat TM data. The USGS is also more than halfway through a major effort to map existing land cover for the United States, at approximately 100 m resolution, also using Landsat TM data. Several national- and continental-scale data sets—acquired by federal agencies over the past few years to promote land cover studies for the United States, the humid tropics, and North American boreal forests—are now available to the scientific community for analysis. At a global scale, the first complete 1 km resolution global land-cover product was released by the International Geosphere Biosphere Program (IGBP) in Summer 1997.

A land *use* indicator will need to distinguish among forms of a general category of land use depending on the criteria by which they are managed. For example, forest lands should be segregated into categories such as primary forest, unmanaged second-growth forest, forest managed primarily for timber production (referred to here as *timberland*), recently burned forest, forest managed for biodiversity preservation (e.g., forests in Safe Harbor agreements), and forest reserves, as well as into categories such as deciduous forest, boreal forest, and Pacific coast rainforest. Such categories can be aggregated into fewer categories as needed, but to identify forests in which the human impact is being reduced, is not changing, or is increasing, the disaggregated input data will be needed.

As an example, we discuss wetlands in some detail. Categories of aquatic habitats include wetlands, fresh and saline lakes, reservoirs, rivers, and bays. Most difficult to identify and monitor are wetlands—waterlogged landscapes that are inhabited by distinctive biotas (NRC 1995b). Wetlands cover 26 percent more area in the coterminous United States than all other categories of aquatic habitat combined (Frayer 1991). Their biotic communities respond chiefly to the influence of hydrology, driven by topography and climate, but also to nutrient supply related to geology and soils. In many locations, because of anaerobic conditions, wetlands accumulate substantial deposits of organic detritus. Wetlands are called peatlands when accumulations of partly decomposed organic matter reach a depth of 30 cm. Many diverse oxidation/reduction reactions mediate elemental fluxes between the atmosphere and wetlands (Mitsch and Gosselink 1993). Wetlands are of major significance for the cycles of carbon, nitrogen, and sulfur, and peatlands are a reservoir of at least 400 billion tons of carbon worldwide (Woodwell et al. 1995). Significant losses of carbon from that reservoir are projected if the global climate warms (Gorham 1991, 1995a). Therefore, monitoring the extent and status of various categories of wetlands is extremely important.

Many physical, chemical, and biological criteria have been used to categorize and monitor wetlands (Adamus and Brandt 1990, NRC 1995b), and a land use indicator will need to recognize a substantial number of wetland types. As in the case of terrestrial land use types, specific categories of wetlands should be separated according to the criteria by which they are managed. A particularly important category is wetlands created as part of mitigation settlements under Section 404 of the Clean Water Act (Kusler and Kentula 1989), because created wetlands seldom fully replace naturally functioning wetlands (Erwin 1991, Gorham 1995b, NRC 1995b, Race and Fonseca 1996). It is also important to recognize as a separate category wetlands restored after years of tile drainage and crop production under the government's Wetlands Reserve Program, which since 1990 has been administered by the Natural Resource Conservation Service (formerly the Soil Conservation Service) under the Food and Security Act of 1985.

Wetland types can be distinguished using data from the National Wetland Inventory (NWI) carried out by the U.S. Fish and Wildlife Service since the mid-1970s, mostly by means of 1:60,000 color-infrared photography. By late 1991, 70 percent of the coterminous states and 22 percent of Alaska had been mapped (NRC 1995b), the basic mapping units being the set of wetland categories devised by Cowardin et al. (1979). The NWI has prepared a report on wetland status as of the mid-1980s, and on the trend of losses since the mid-1970s (Dahl and Johnson 1991), based on a representative sample of U.S. wetlands. Future reports are planned at 10 year intervals. Other NWI products include reports to accompany each 1:100,000 scale wetland map, reports on state wetlands, and a wetland-plant database (Mitsch and Gosselink 1993). NWI maps are being digitized for a computerized geographic information system that will facilitate both analysis and display of the data. In only ten states, however, was the process near completion in 1994 (NRC 1995b).

Reliability. Calculating changes in the proportions of land cover in the United States with fully replicable techniques will require integrating remotely sensed information with data from statistically based, in situ sampling programs. A growing number of examples exist in which such information has been collected on regional scales, especially using remote sensing. Of particular interest to the scientific community are studies of tropical deforestation (Skole and Tucker 1993; Laporte et al. 1995; Janetos et al. 1997 [GOFC]), in which changes in proportions of land cover types have been calculated and full transition-probability matrices have been derived.

Temporal and Spatial Variability. The land cover indicator itself does not retain the spatial information in the underlying data. However, if the underlying data are archived with all their spatial information intact, then more complex, spatially explicit measures of land cover changes can be computed (Appendix B). These measures are likely to be of substantial regional and local interest.

Many natural processes cause changes in land cover, but typically they result in relatively small annual changes. Appreciable changes typically occur only over decades. However, some anthropogenic changes, such as clearing forests for rangeland or farmland, and large fires, happen quickly. Therefore, land cover needs to be computed annually but it need not be formally reported more often than once every five years.

Statistical Properties. The statistical properties of land cover are clearly a function of the reliability of identification of land cover types from remote sensing and of the statistics of the in-situ sampling program. Because the use of remote sensing to quantify land cover change is an active area of research and application within the scientific community and operational agencies, the statistical properties of land cover should be clarified in the near future.

The classification of land cover types in the land cover indicator must be sufficiently comprehensive for all the additional indicators that are derived from it. Classification issues can be addressed in two ways. One would be to determine the number of classes required by the most demanding indicator and use that classification in all input data. This method would force the unification of classification schemes for all indicators. The other would be to have a more flexible, hierarchical classification scheme that would allow the unfolding of additional classes of land cover from a smaller number of aggregated classes from the same underlying data. The latter approach has been used by the IGBP (IGBP 1992b; Defries and Townshend 1994; Townshend et al. 1994) in its global 1 km land cover product. A flexible, hierarchical classification system would best serve the needs of a variety of environmental indicators.

The rate of land cover change determines the most appropriate sampling intervals, but there are also practical limitations on sampling frequencies. For example, analyses of satellite data are constrained by both the technical features of the satellite itself and by the ability of investigators to handle and interpret the very large volume of data. In addition, the return time for sampling NRI plots or CFI plots is largely determined by the availability of field crews. However, the mean resampling time for CFI plots is approximately 10 years, which compares favorably with theoretically determined appropriate sampling frequencies for forest productivity (Appendix A).

Scientists who use remote sensing have addressed issues of sampling intervals at some length (Skole et al. 1997; IGBP 1992b; Justice and Townshend 1988). As a rule of thumb, they have suggested that five-year intervals of complete remote-sensing surveys at national, continental, and global scales are generally adequate. Complete surveys could be interspersed with annual stratified samples to detect rapid changes without overwhelming the capacities of investigators and systems to perform the analyses.

Necessary Skills. The collection of remotely sensed data obviously demands a high degree of familiarity and experience with the instruments and data-handling capabilities. However, technological barriers to handling remotely sensed data are shrinking as computer technology improves and costs decline. The greatest barriers to developing and using the land cover indicator are probably conceptual: developing the detailed techniques for classifying, combining, and interpreting changes in land cover categories by means of which remote-sensing and in situ data are evaluated.

Data Quality Control, Archiving, and Access. The input data for land cover should be archived at the most highly resolved and disaggregated levels, and the techniques used to generate land cover classes need to be described clearly and documented. Only in this way will the land cover indicator be replicable and real changes detectable. Sources of error in both measurements and classifications should also be clearly defined and documented. Comparing maps derived at different times, although feasible, is not a good method of documenting changes in land cover because it confounds measurement errors, interpretation errors, and cartographic errors in ways that would be extremely difficult to quantify. It is more straightforward and desirable to detect change in the underlying data themselves, use those differences for quantitative analyses, and then derive maps for presentation purposes.

Robustness. Although tremendous strides have been made in recent years, the use of remotely sensed information for ecological analyses is still in its infancy. For many years, the technical challenges of simply handling and processing the data were so large that they inhibited the use of the systems by all but the most sophisticated laboratories. Rapid improvements in cost and performance of computer hardware and software are removing many of these technical impediments, but other issues remain. The most important ones include the care, maintenance, and accessibility of data archives, and the intercalibration of the remote-sensing instruments themselves.

The land cover indicator is likely to be robust to reasonable sources of interference, especially if the original data are archived carefully. However, the time series of measurements can be compromised by technological changes unless sufficient care is taken to ensure that new instruments are cross-calibrated, and that the calibration of instruments is maintained and monitored carefully over their lifetimes. Achieving calibration precise enough for these quantitative scientific measurements is difficult, but it can be achieved, as the Landsat data record shows. For data sets that last longer than the lifetime of any one instrument, successive instruments must be flown and cross-calibrated for a period of overlap. In this way, trends due to degradation of instrument performance and offsets due to changing satellite orbital geometry or new technology can be identified and quantified before additional analyses are performed.

International Compatibility. International programs within the IGBP, the Committee of Earth Observing Satellites (CEOS), and the Global Terrestrial Observing System (GTOS) have made similar recommendations. There is also substantial international activity through programs designed to use large amounts of remotely sensed and other information to track changes in land use experience (IGBP 1992a; IGBP 1995; Janetos et al. 1997; Justice et al. 1993; Justice et al. in press; Kirchoff 1994; Turner et al. 1993). In principle, the land cover indicator could be derived similarly from any of these other international efforts. It will certainly benefit from all of them.

INDICATORS OF ECOLOGICAL CAPITAL: BIOTIC RAW MATERIALS

The United States has repeatedly affirmed its commitment to preserving its biological resources (see for example the Endangered Species Act [ESA] of 1973 and its various amendments). Currently, the only accounting of our success (or failure) is the number of species listed as endangered or threatened on the ESA Endangered Species List. This list contains only species for which there is some minimal amount of information and interest, and it is influenced by political and economic factors, as well as by information about biology and the status of populations. For these reasons, the list is not an accurate reflection of the number of species at risk of extinction, or, more generally, of how well the nation's biological resources are faring.

The committee recommends two indicators of ecological capital. The first, *total species diversity*, is a measure of the ecological capital actually present. The second, *native species diversity*, compares the number of native species an area of land supports with the number it would support in the absence of human impacts. Two other indicators, which measure the original ecological capital that remains after human impacts, and the ecological capital borrowed from somewhere else, are proposed in Chapter 5, because they are most useful as indicators at local and regional scales. Gathering the data needed to compute these indicators will require substantial investments of human and financial resources, but a major effort is needed to achieve the nation's commitment to preserving its biological resources. Both indicators depend ideally on a land use indicator, but while that indicator is being developed, these should be developed using the information available, whether via land use or land cover indicators.

Total Species Diversity

The simplest measure of species diversity is species richness, an unweighted list of the species present in any unit of land. Because loss of a species is irreversible, species richness is especially important to monitor. Weighted indices of species diversity are valuable for many purposes, but because they usually discount rare species, which are often our primary concern, we recommend indicators based on unweighted measures (species richness). The land use indicator, when it is developed, can be used to build an indicator of total species diversity by assigning a score to each category of land use, representing its contribution to preserving species, and then computing the average score for the nation as a whole. That average—the total species diversity indicator—can be computed by multiplying each score by the number of square kilometers in its land use category, summing scores, and dividing the total by the number of square kilometers in the nation. Until that indicator is developed, *land cover* should be used in its place.

Assigning scores for each land cover category is the difficult problem. The simplest way would be to use the number of species in each land cover category as the score. The total species diversity indicator would then be the average number of species in all land cover categories. However, this method fails to consider differences in the areas covered by each land cover category. The number of species in an area depends on its extent; larger areas have more species than smaller ones (Arrhenius 1921).

For example, if a decision to recognize a new land cover category is made, a former category will need to be split. Because each part is smaller, each will have fewer species than the old category. As a result, the total species diversity indicator would decline without a change in the status of any species. Therefore, any useful indicator must adjust for area differences. Often, the value is normalized by dividing the diversity by the area to obtain a new measure of the "density of species." However, this new measure is inappropriate because diversity does not increase linearly with area. Ecological experience for almost two centuries has shown that, within a continent or biogeographic province, the relationship between land area and total biological diversity fits a power law called the species-area curve:

$$S = cA^z,$$

where S is the number of species, A is area, and c and z are coefficients of the equation (Arrhenius 1921, Preston 1962).

If z were 1, then diversity would be linearly related to area and no adjustment would be necessary. But z is close to 1 only when similar biological provinces are compared (such as North America and Eurasia). Typically z is about 0.15 for different areas of a single land cover category in a single continent (Rosenzweig 1995). That is, an area 10 times larger than another will not have 10 times the species diversity of the smaller, but only 10^z times as many species. Using the typical z-value of 0.15, $10^{0.15}$ is only 1.4. Dividing S by A would make the larger area appear to have a density only 14 percent as high as that of the smaller area, despite both areas' being samples of the same whole and thus having the same intrinsic species densities.

The power law is neither precise nor entirely accurate (Leitner and Rosenzweig 1997), but it gives sufficiently good fits in a wide variety of circumstances to provide the basis for a quantitatively defensible adjustment for area.

$$D_i = S_i/A_i^z,$$

where D_i is the adjusted species density of land cover type i. In other words, divide the number of species, S, by a function of area, A^z, because A^z *is* linearly related to diversity. If a land cover category sampled at two different scales is adjusted in this manner, its species density will be the same. To compute the total species diversity indicator, multiply each D_i times p_i, the proportion of i in the land cover indicator. The sum of these products would then be an estimate of total species diversity, although this estimate needs to be related to some reference state, namely, the amount of biological diversity expected in that land cover type. The power law can be used for this purpose as well.

For example, to determine the expected wildflower diversity in northwestern timberland, one first measures the wildflower species-area relationship on northwestern forest reserves (i.e., places set aside as parks and wildernesses that can serve as standards for the natural ecosystem). This yields the reference values of c and z. Second, the number of wildflower species are measured in an average square kilometer of timberland, S_{tmbr}. Substituting 1 km^2 for A, and the values of c and z obtained from the forest reserves, the referent or standard S_n for the square kilometer of timberland can then be calculated. S_{tmbr} can be compared with S_n using a simple ratio, S_{tmbr}/S_n. However, doing so would imply that the more species, the better the state of diversity. Instead, unusually high diversity, that is, values of S_{tmbr}/S_n greater than 1, are likely to presage a decline of diversity. To correct for this problem, the score assigned to the category "timberland" should be based on the absolute value of the difference between it and a forest in reserve. This value can be standardized by dividing it by S_n. For ease of interpretation, the new measure can be scaled by subtracting it from 1:

$$M_{\text{tmbr}} - 1 - \{ |S_{\text{tmbr}} - S_{\text{n}}| / S_{\text{n}} \}.$$

This scoring standard has a maximum value—a perfect M—of 1. Mathematically, M can drop below 0 if a land cover category has more than twice S_n, but it would fall to 0 only in a sterilized environment. Therefore, M can be treated as if it were a true proportion.

The assessment can be repeated with a second taxon, such as butterfly species, a third, such as birds, and a fourth such as phytoplankton in lakes. In principle, there is no limit to the number of taxa that can be used. However, for many taxa, lack of data or taxonomic knowledge prevent their use at present. The quickest, cheapest, and most readily interpretable scores will be achieved if the diversity surveys are limited to a small number of easy-to-survey taxa with high aesthetic or recreational value to Americans. The final score will be the average of the separate scores from the different taxa.

In aquatic ecosystems, similar procedures can be used to calculate M. Reference lakes and streams are needed to establish the expected numbers of species in different taxa, such as phytoplankton, zooplankton, periphyton, insects, and fishes.

To compute total species diversity, this process is repeated in each category of terrestrial and aquatic land cover, to calculate a separate score for each. Then each score is multiplied by the proportion of the nation devoted to that particular land cover, $M_i \times p_i$ (where p_i is the proportion of i in the land cover indicator). Total species diversity, which is the sum of these products, will be a proportion with values from 0 to 1. It would give the nation an overall view of its biodiversity. The indicator's chief value is in providing a measure of total species richness. It can reflect human impacts, especially severe ones, and it also reflects many other environmental variations. Thus it allows one to compare the species richness in various land cover types as well as the effects on species richness of various natural environmental and human-caused changes.

Because species counts are the backbone of any ecological survey, lists of species are available for many areas. The areas sampled need not be the same size, because, using the power law, they can all be adjusted to a fixed area.

Once appropriate land cover categories are determined for the land cover indicator, the referent standard species-area relationship for each land cover category's natural ecosystem is readily calculated, if large enough samples are available. Data from very small samples fit the power law poorly; they badly underestimate diversity. Underestimates erroneously reduce the estimate of c and increase the estimate of z, changing S_n. The errors in estimating c and z tend to offset each other, but they do so to an extent that depends on the particulars of each case. Hence, it is best to avoid small samples as much as possible.

The scale at which *M* is measured also affects its value. Some native species are rarely and irregularly found in a small proportion of a land cover type. Such species are neither typical nor sustainable under these conditions. They actually depend on their native, natural ecosystems. Nevertheless, if one counts all the native species in the entire area devoted to a land cover category, rare species will be scored. This is a problem because the indicator *M* depends on simple counts of species, and so it counts a rare species found in only a small number of samples exactly the same as a species with a large, widespread, healthy population in that land cover category.

It would be desirable to solve this problem by replacing the number of species, *S*, with a diversity index that takes abundance into account (e.g., Simpson's index or the Shannon-Weiner index), but not enough is known about how species abundances are distributed in natural and disturbed ecosystems, or how diversity indices behave as functions of area to do so. Therefore, replacing *S* with a weighted index would not improve the indicator. As knowledge of abundance patterns improves, use of weighted indices might become practical. Meanwhile, an appropriate adjustment for rare species is to estimate *M* using many subsamples of the land cover category. Each subsample should be large enough to have a good proportion of the native species, but small enough so that rare and unstable populations are recorded in only a few of the subsamples. The few subsamples that have such rare populations will be appropriately diluted in the average *M* of the set of subsamples.

When calculating *M*, land cover types should not be aggregated into a single category. To see why, consider streams and their fishes. Ignoring the effect of aridity on the area of freshwater ecosystems, there is a slightly negative (but nonsignificant) species-area curve among the native species found in each state. This relationship occurs because big states tend to be arid or, in the case of Alaska, cold and with low ecosystem productivity. Because arid areas have fewer and smaller streams, the area of a state is a poor measure of the extent of its fresh waters. Our expectation of fish diversity, our referent state, needs to be computed from subsamples whose spatial variation in aridity and productivity fall within narrow enough bounds to keep them in the same climatic zone.

If the effects of human activity on the environment continue to increase, many species are likely to become extinct. If the referent standard values of S_n are adjusted to that new situation, a steady decline in S_n and a concomitant steady but fallacious rise in total species diversity would result. Recalculating S_n would be an unfortunate example of the hazards of the shifting baseline phenomenon (Pauly 1995). S_n needs to stand as a fixed standard for each sort of broadly defined ecosystem. Once measured well, it should not be changed even if diversity declines in protected ecosystems. Only if the standard is maintained will the United States have a sound measure of how well it is doing in maintaining its biological capital.

Native Species Diversity

This indicator reflects human impact on the land. Land that has been so transformed by people that it cannot support most of the native species that would otherwise be there is land that carries a heavy burden caused by human activities. In contrast, the human impact on lands that still support a diverse assemblage of native species is light. We call our indicator the *Native Species Diversity* indicator. The *total* species diversity indicator counts all the species of a taxon—native and exotic. The *native* species diversity indicator covers only natives, because its purpose is to measure human impacts. If humans cause a native species to be replaced by an exotic, native species diversity counts that as an impact; total species diversity does not.

Human population pressure is the basic cause of reductions in environmental quality, including loss of native species, but population size is not the sole determinant of those losses. Many damaging practices, such as release of toxic materials and overexploitation of renewable resources, have accompanied even low-density human populations. Therefore, human population size itself is not a good environmental indicator, even though environmental threats would be fewer if there were fewer people.

The human population has environmental impacts by converting natural ecosystems into places to live and work, although dwelling places, shopping centers, office buildings, factories, and public utilities constitute only a small part of the impact. In the United States, only 35 percent of the total land area impervious to water is covered by places for people to live and work; the other 65 percent is related to transportation, primarily roads and parking spaces for cars. The total area of the United States that is impervious (e.g., roofs, concrete, and asphalt surfaces) is still a small percentage of the total land area, but it has doubled since World War II.

In general, when the fraction of impervious area increases above 10-20 percent of the total land in a given area (e.g., a watershed), hydrologic flow patterns change markedly from natural conditions and diffuse source water pollution problems tend to occur. In addition, much larger areas are used to grow the food, medicines, and construction materials that sustain health and well-being. These materials are also transported over distances that would have amazed people a century ago. For example, Folke et al. (1996) estimated that the 29 cities of the Baltic region appropriate natural resources from ecosystems that cover 200 times the total area of the cities themselves.

Several statistics illustrate cogently the need for national assessment of our impact on the land. First, less than 1 percent of U.S. grassland remains in anything like its natural state. Second, only 4.7 percent of the nation's forest land is unmanaged, even though most formerly forested land still has trees growing on it. Third, human activities, primarily agricultural drainage, resulted in the loss by 1970 of more than half of the wetlands present in the United States at the time of European settlement (NRC 1995b). In some states, for example, Ohio and Iowa, the loss exceeds 95 percent. Substantial losses have continued since then.

Not all conversions of land to our own direct use exterminate the species that live there, although few wetlands species can survive drainage of their habitat. The magnitude of the effect depends on the wisdom exercised in managing lands. Therefore, indicators are needed that measure the amount of land converted to different purposes and how well the converted land remains productive and supports biological diversity.

Description of the Indicator. As in total species diversity, species-area power laws supply the standard to evaluate the human impact. Native species diversity uses the same power-law relationships that were measured in natural reserves for total species diversity ($S_n = cA^z$), but thereafter its construction differs. To understand the difference, consider again the example of scoring northwestern timberland on the basis of its native wildflower diversity. To calculate total species diversity values of c and z, and obtain the natural number of species, S_n, for a square kilometer, all species were counted. To calculate native species diversity, exotic species are excluded from S_{tmbr}. The remainder is the native wildflower diversity, $S_{n,tmbr}$. The score for human impact is the proportion of that standard achieved in the timberland:

$$G_{tmbr} = S_{n,tmbr} / S_n.$$

Because $S_{n,tmbr}$ cannot exceed S_{tmbr}, G_{tmbr} is a proportion, with values ranging from 0 to 1. On pristine land, $G = 1$; a land cover category with no native species receives a score of 0.

People manage forestlands and other lands with a variety of strategies, some designed to protect the environment, others not. To incorporate management variation into its value, G is measured separately for lands managed by different strategies and then multiplied by the proportion of the land type subject to that strategy. The sum of these products over all strategies is the score for the land category. It, too, is a proportion, with values ranging from 0 to 1.

To illustrate the calculation of G, consider Cody's (1975) determination of the species-area power relationship for birds in California chaparral. The natural power relationship is

$$S_n = 45A^{0.125},$$

To evaluate G for residential areas around a city, breeding bird lists are obtained and the average number of native species (S_i) enumerated for a set of square kilometer samples of residential land in what would otherwise have been chaparral. The power law tells us to expect 45 species in each square kilometer, so $G = S_i/45$. Because of the species-area power law, one is not restricted to areas of 1 km^2; data from a variety of sample areas can be used. (For an example from Tucson, Arizona, see the discussion of biodiversity indicators in Chapter 5.)

A similar measure commonly used as an indicator of aquatic ecosystem condition is based on concepts used by Karr et al. (1986) in developing the Index of Biotic Integrity (IBI). The IBI is an additive index that has been locally calibrated; users need to have only sufficient knowledge to be able to identify local species. Multifactorial indicators such as IBI have been developed around a set of measures of the distributions and relative abundances of selected taxa. Each factor is assigned a numerical value (an integer between 0 and 6) based on the professional judgment of the evaluator. The assignment of the appropriate integer value is based on the distribution of the data for each factor from a number of reference sites. The final indicator is calculated as the sum of the individual factor scores (usually 10 to 12), which typically generates a score between 0 and 60. Indicator developers then set management goals for a particular water body based on a predetermined ranking score for the indicator.

Although IBIs have been helpful in the development and evaluation of management policies in many regions of the country, the subjective nature of the judgments in assigning values to each factor, as well as problems associated with calibrating multifactorial indices (Reynoldson et al. 1997), lead the committee to recommend native species diversity rather than IBI as the best national-level indicator of human impact on species diversity. However, IBIs have considerable value at local and regional scales and we encourage their continued use.

Changes observed in native species diversity come from several sources. Native species diversity decreases if more land is shifted to human use and if human use is intensified without regard to the ecological consequences. Native species diversity increases if management strategies on some lands are changed from those with heavy impacts to those with lighter impacts, if management strategies improve, and if land shifts to a less ecologically damaging use.

An important case of an improved land use strategy is the Safe Harbor program, an example of cooperation between private landowners, a nongovernmental conservation organization (the Environmental Defense Fund), and the federal government. In return for landowners' managing their land to encourage its colonization by endangered species, landowners become exempt from the restrictions that the U.S. Fish and Wildlife Service would otherwise place on their property were an endangered species to move in on its own.

Land in the Safe Harbor program should be separated from other similar land into a special land use category in the land cover indicator. It would have an entry in the land cover indicator and special measurements would be made of its contribution to total species diversity and native species diversity. Native species diversity provides an indication of the success of conservation over the entire slate of land uses and management strategies.

Although lack of adequate information on many taxa will make developing these indicators of species diversity difficult, the work should be started now. Doing so will provide an incentive to learn about taxa that are not well known at present. There is enough information to be useful already and the rate at which our environment is changing makes it urgent to implement such indicators.

INDICATORS OF ECOLOGICAL CAPITAL: ABIOTIC RAW MATERIALS

Nutrient Runoff

Water quality and ecological conditions in U.S. coastal waters have been subjects of growing national concern for more than two decades. Several major types of pollutants—oxygen-demanding organic matter, microbial pathogens, heavy metals, synthetic organic compounds that bioaccumulate to potentially toxic levels, and excessive levels of nutrients—all have taken their toll on these ecosystems. Massive efforts sparked by the Clean Water Act to clean up municipal and industrial sewage effluents have yielded substantial reductions in point-source pollution and improved water quality near such sources. Efforts to control potentially toxic heavy metals and synthetic organic chemicals have resulted in major declines in loadings of these substances to coastal waters, but the legacy of many years of inadequate controls is still seen in the polluted sediments and high body burdens of these substances in the marine organisms of many coastal areas. Among the major potential pollutants affecting coastal environments, nonpoint sources of nutrients—N and P, in particular—have received relatively little regulatory attention.

The results of excessive nutrient loadings are seen in reduced water clarity; nuisance algal blooms, including species with toxic forms like *Pfiesteria piscicida* and *Gymnodinium* spp. (red tide); and hypoxic (low oxygen) bottom waters. Outbreaks of toxic and other algae have been correlated with nutrient enrichments and appear to be increasing in estuarine and coastal waters (Burkholder 1998). Hypoxia, generally defined as persistent oxygen concentrations of less than 2 mg/L, affects areas of Long Island Sound, Chesapeake Bay, the near-shore Gulf of Mexico near the mouth of the Mississippi (e.g., Rabalais et al. 1996), and many other coastal areas around the world. In the Gulf of Mexico, the affected area has grown from about 9,000 km^2 in 1985 to approximately 18,000 km^2 at present. A large increase in the affected area occurred in 1993, apparently in response to the large spring flood, which brought a corresponding increase in nutrients into the Gulf from the Mississippi River. Current evidence links hypoxia primarily to increased inputs of nitrogen forms (mainly nitrate) to coastal waters, inputs that stimulate algal growth. Nitrogen generally limits plant growth in coastal and ocean waters, whereas phosphorus is usually the limiting nutrient in fresh waters. Although growing algae produce oxygen in surface waters, decomposition of dead algae in bottom waters consumes oxygen, leading to loss of habitat for fish and other forms of aquatic life.

Human alterations of the biogeochemical cycles of major nutrient elements have reached global proportions. For example, human contributions to the cycling of nitrogen forms equal the contributions from all natural processes (Ayres et al. 1994). Although nearly all human additions to nutrient cycles occur in terrestrial ecosystems, these systems are leaky, which is why nutrient loadings to coastal zones are elevated substantially above background levels continent-wide. Quantitative data on these increases are sparse, however, and insufficient to document temporal trends in nutrient losses from the continental United States to coastal and marine systems. Evidence is also lacking on the geographic extent of impacts of elevated nutrient loading on the open oceans, although it is generally assumed that impacts of human-induced nutrient inputs on the productivity of the oceans as a whole are still negligible.

Because of widespread concern about the impacts of nutrient loadings on coastal waters, and the lack of quantitative information to develop related national policies, the committee recommends the development of national- and regional-scale indicators for N and P runoff from the land to coastal areas. Data are already being collected to produce such statistics. The USGS monitors flow rates of major rivers, and state and federal agencies gather water-quality data routinely at stations near the mouths of most rivers.

Obtaining accurate loading estimates from routine monitoring data is difficult. Flow rates and nutrient concentrations vary substantially in space and time; measurements of flow and nutrient concentrations are often not made simultaneously; concentrations are usually measured infrequently relative to the time scale of their variability; and sampling networks have spatial biases because of the need to target sampling toward specific pollution sources (Smith et al. 1997). However, models can be used to estimate nutrient fluxes from such data (e.g., Smith et al. 1997). Current data appear adequate to characterize the yields of nutrients from the major hydrological cataloging units in the country (Figure 4.1) and to estimate total runoff from the conterminous United States to coastal waters.

An even larger effort has estimated N and P fluxes to the North Atlantic Ocean from rivers in 14 regions of North and South America, Europe, and Africa (Howarth et al. 1996). According to these authors, nonpoint sources of nitrogen dominate riverine fluxes to coastal waters in all regions. On an areal basis, the largest N fluxes are from watersheds in northwestern Europe and the northeastern United States (Figure 4.2); but on a mass basis, the Mississippi River drainage basin is by far the largest

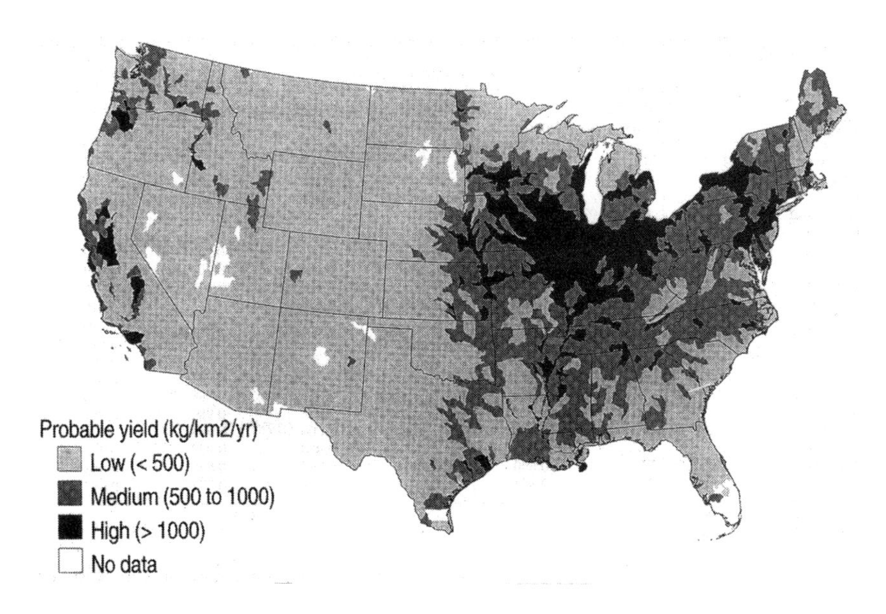

FIGURE 4.1 Classification of predicted local total nitrogen yield in hydrologic cataloging units of the conterminous United States. Local yield refers to transport per unit area at the outflow of the unit due to nitrogen sources within the unit, independent of upstream sources. Source: Smith et al. 1997. Reprinted with permission from the American Geophysical Union.

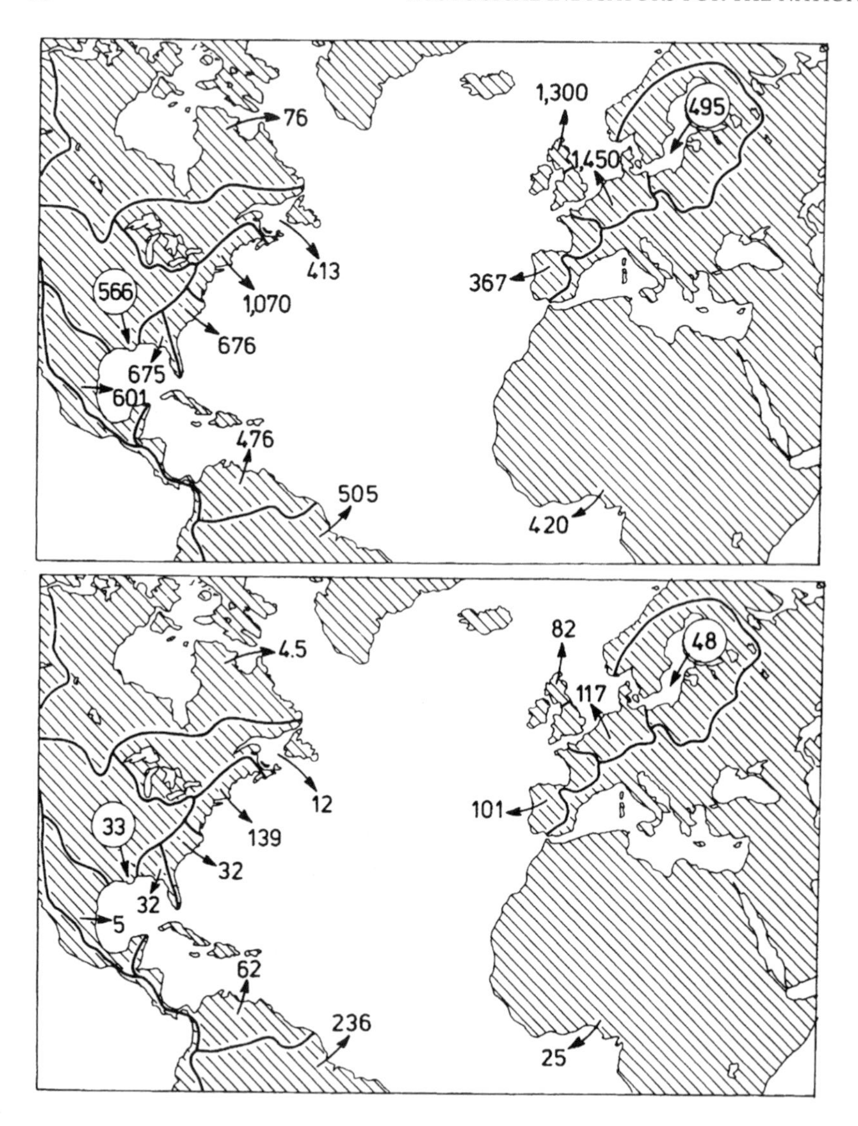

FIGURE 4.2 Total nitrogen (TN) runoff in rivers in kg N km^{-2} yr^{-1} (top) and total phosphorus (TP) runoff in rivers in kg P km-2 yr^{-1} (bottom). See text for explanation of estimates. Source: Howarth et al. 1997. Reprinted with permission from Kluwer Academic Publishers.

single contributor of both N and P to the Atlantic Ocean from North America (1.82 of 4.61 teragrams [10^{12} g] per year [Tg/year] for N, and 0.107 out of 0.241 Tg/year for P). A strong linear correlation exists between river fluxes of total N and the sum of anthropogenic N inputs to temperate regions, but on average, regional fluxes in rivers are only about 25 percent of the computed anthropogenic inputs. Based on data from relatively pristine areas, Howarth et al. (1996) estimate that riverine nitrogen fluxes in many temperate regions have increased 2- to 20-fold from preindustrial times. Like the previous study, this study did not produce runoff-rate statistics for any particular year. Instead, both used information gathered over multiple years to produce annualized runoff rates generally representative of "current" conditions.

The committee's recommended indicator of total runoff of N and P can take values ranging from 0 (no discharge) to thousands of kg km^{-2} yr^{-1}, with lower values being more desirable for most purposes. The status and trends in the total runoff of N and P should be computed and reported annually (e.g., in Tg/year). Annual rates for major river systems such as the Mississippi River should be reported together with aggregate numbers. Because of the strong influence of short-term weather variations (drought or flood conditions) on regional runoff rates, the annual output runoff data (both national and regional) should also be normalized by dividing the runoff rate (Tg/yr) by the ratio of actual rainfall over the region to the long-term average rainfall for the region. Normalized data provide insights into the sensitivity of river basins to short-term weather variability.

On a time scale of a few decades (or less), the national indicators of N and P runoff to coastal waters can be evaluated and used in determining the effects of national policies to control point and diffuse sources of nutrient pollution on the net loss of nutrients from the country. On regional scales, the statistics can play similar roles in evaluating the effectiveness of specific regional management practices. Comparison of trends in runoff with trends in fertilizer use, crop production, and coastal water quality can give further insights into the need for changes in policies, regulations, and management practices.

Soil Organic Matter

Soils promote the growth of vegetation, including crops; control the flow paths of precipitation as it becomes surface and groundwater; and serve as a filter for potentially harmful substances that would otherwise enter this water and the atmosphere (Larson and Pierce 1991, Parr et al. 1992, Johnson and Lindberg 1992, NRC 1993). The ability of soils to perform those functions depends on their physical, biological, and chemi-

TABLE 4.2 Reference and Measured Values of Minimum Data Set for a Hypothetical Typical Soil Type (Hapludoll) from North-Central United States

Horizon and Characteristic	Reference Value	Measured Value
Surface Horizon		
Phosphorus (mg/kg)	30	15
Potassium (mg/kg)	300	300
Organic carbon (percent)		
Total	2	1.5
Labile	0.2	0.15
Bulk density (mg/m^3)	1.3	1.5
pH	6.0	5.5
Electrical conductivity (S/m)	0.10	1.0
Texture (percent clay)	30	32
Subsoil horizon		
Texture (percent clay)	35	35
Depth of root zone (m)	1.0	0.95
Bulk density (mg/m^3)	1.5	1.5
pH	5.5	5.5
Electrical conductivity (S/m)	0.10	0.10

Source: NRC 1993.

cal properties, which are easily measurable and whose relationships to these functions are well known. The NRC (1993) reviewed the potential for soil properties to serve as indicators of the soil's ability to promote plant growth, regulate water flow, and filter or retain chemicals of concern. Its summary of a minimum data set is given in Table 4.2. The NRC study (1993) identified soil organic matter (SOM) as "perhaps the single most important indicator of soil quality and productivity," and work by Bauer and Black (1994) and Reeves (1997) also supports the use of that indicator. Therefore, the committee recommends soil organic matter content as the best currently available indicator of the state of soil quality. SOM is an indicator of ecological condition (soil condition, relationship to erosion) and of ecological functioning (soil productivity).

The Indicator. Soil organic matter strongly influences several biological, physical, and chemical characteristics of soils. SOM is a nutrient and energy source for soil biota; it improves soil structure by strengthening soil aggregates, increases water retention and available water capacity, reduces the sealing of soil surfaces thereby promoting infiltration and reducing erosion, increases cation exchange capacity, chelates metals, and influences the fate of pesticides. SOM responds to tillage and fertilization and it is sensitive to erosion, as the uppermost portion of the soil normally has the highest organic matter concentration. Soils high in organic matter are generally productive, and the carbon content of soils (approximately half of the SOM weight) is an important component of national and global carbon budgets. Perhaps the biggest disadvantage of SOM as an indicator is that it is likely to change slowly, and real trends in SOM are difficult to detect because SOM values are highly variable at both local and regional scales.

Numerical Values of the Indicator. The five major factors that affect soil formation—climate, organisms, topographic relief, parent material, and time—control the natural distribution of SOM. Regional differences in the amount and distribution of SOM with soil depth are pronounced, as are differences in those constituents across landscapes with strong gradients in soil-drainage class (e.g., depth of the water table). Concentrations of SOM in agricultural soils generally range from 1 to 10 percent in surface horizons, to less than 1 percent at greater depths. Long-term tillage of agricultural soils has reduced SOM content (Jenny 1941, Paustian et al. 1997), probably as the result of erosion, increased soil temperature, and reduced organic matter input. SOM can recover or be maintained through careful management (NRC 1993, Reeves 1997).

Data Requirements. Data are not currently available to compute widespread, accurate baseline estimates of SOM. Therefore, for the moment, the SOM content of a system should be expressed as a net or percent change over time. Three methods are used to express SOM quantities: gravimetric, volumetric, and equivalent mass. The gravimetric basis, the ratio of SOM mass to soil mass, does not account for soil density. The volumetric basis, the ratio of SOM mass to soil volume, does not account for the total amount of soil present. The equivalent mass basis, the mass of SOM in a standard mass of soil, corrects the volumetric basis to account for total soil mass. The volumetric and equivalent mass bases require soil density estimates. Soil density, particularly of surface soils, varies over short periods, but equivalent mass measures are not sensitive to soil-density changes. Therefore, at this time, the SOM content of a soil is best expressed on an equivalent mass basis.

SOM data can be obtained from the soil samples routinely sent to agricultural soil-testing laboratories using well-established, inexpensive measurements or through a specially devised random sampling program. Protocols for SOM measurements exist (ASA Methods Manual), and the measurements are easy to make. Currently, samples taken typically include only the top 20 cm, or the area subject to plowing, the zone most likely to respond to management practices. With bulk density samples of the top 20 cm, SOM mass per unit area can be calculated to a 20 cm depth. Nevertheless, in many soils, such as conventionally tilled soils, there are significant amounts of SOM below 20 cm. In a program with sufficient personnel, SOM mass per unit area determined to a depth of 50 cm would be preferable.

INDICATORS OF THE PERFORMANCE OF THE NATION'S ECOSYSTEMS

The performance of the nation's ecosystems, or their ecological functioning, can be measured by their productivity, the rate at which they use (mainly solar) energy to fix atmospheric carbon dioxide. The carbon economy of ecosystems can measure the productivity of the nation's ecosystems and estimate their overall carbon budget, that is, whether the ecosystems are losing or accumulating carbon. The committee recommends four indicators of this carbon economy and a related one, dissolved oxygen, that is also related to productivity. The first of these indicators related to productivity, total chlorophyll per unit area (g/m^2), provides a direct measure of the *production capacity* of terrestrial ecosystems (the equivalent measure in lakes is chlorophyll per unit volume). The more chlorophyll per unit area (or volume) in an ecosystem, the greater its capacity to capture sunlight. The second indicator, *net primary production* (NPP) (g carbon/m^2/year; g/m^3/year in aquatic systems), is a direct measure of the amount of energy and carbon brought into an ecosystem. It is also a measure of productivity in the sense commonly used in agriculture and forestry: the amount of plant material produced in an area per year. The third indicator, *carbon storage* is measured by net ecosystem production (NEP), and is a direct measure of the amount of carbon sequestered or released by ecosystems per year. Carbon storage is particularly important in view of recent concerns about greenhouse gas emissions, because the total amount of carbon in the form of CO_2 emitted by a region equals the region's fossil fuel emissions minus its NEP. The fourth indicator, *trophic status,* characterizes primary production in lakes. The trophic status indicator is derived by combining measures of Secchi-disk transparency, total phosphorus, and chlorophyll *a* concentrations (Carlson 1977). The committee recommends use of *dissolved oxygen* as an indicator of the performance of flowing-water ecosystems.

In addition to the above five indicators that are directly related to productivity, soil organic matter and land use, discussed elsewhere in this chapter, are also related to ecosystem functioning. Finally, the committee recommends indicators of agricultural nutrient-use efficiency and overall nutrient balance as important indicators of ecological functioning.

Indicators of Terrestrial Productivity

Except for a negligible contribution from chemoautotrophs, all energy that flows through ecosystems is supplied by green plants during photosynthesis. Energy in the form of light is captured by chlorophyll and converted to chemical energy in the form of organic carbon. Therefore, the amount of chlorophyll present determines the fundamental capacity of an ecosystem to capture energy, its *production capacity*. Although soil water, nutrients, and many other factors may limit the rate of energy capture, plants regulate their chlorophyll concentrations in response to these other limits (Hurtt and Armstrong 1996, and references they cited). Thus, the abundance of chlorophyll is an excellent indicator, because it is strongly correlated with an ecosystem's actual capacity to capture energy, not just its potential capacity.

Terrestrial plants obtain carbon exclusively from atmospheric CO_2. The energy captured by an ecosystem in units of carbon gained is referred to as *gross primary production* (GPP). After capture by chlorophyll, energy is stored in organic molecules. Plant respiration consumes about 40 percent of GPP worldwide (Schlesinger 1997), most of it respiration by leaves, some by fine roots, and still less by stems. Roughly 25 percent of the energy gained is used to construct plant tissues (construction respiration, Foley et al. 1996). When all sources of plant respiration are subtracted from GPP, the remainder is *net primary production* (NPP). Because NPP, the amount of organic carbon actually made available to other organisms in an ecosystem, fuels all ecosystem functioning, it is a useful indicator of the carbon budgets of ecosystems.

In many natural terrestrial ecosystems of North America, only 4 to 10 percent of NPP is typically consumed by animals (Whittaker and Likens 1973). Animals consume a much larger fraction of NPP in rangelands, grasslands, savannas, and in freshwater and marine ecosystems (up to 67 percent), environments in which primary production is used to synthesize easily digested tissues, than in forested ecosystems, where much NPP is allocated to the production of wood. NPP that is not consumed by animals eventually enters the soil or aquatic sediments as undecomposed organic matter, where it is metabolized by detritivores and returned to the atmosphere as CO_2. The rate of decomposition is affected by many biological and physical factors, including temperature, pH, soil moisture, nitrogen availability, and the lignin content of the detritus (Parton et al. 1988).

The difference between the sum of all nonplant respiration in an ecosystem (all the CO_2 carbon produced by detritivores and animals) and NPP is *carbon storage* or net ecosystem production (NEP). NEP is the change in the total amount of carbon in an ecosystem. If NPP is less than total nonplant respiration, NEP is negative and the ecosystem loses carbon, most likely to the atmosphere as the greenhouse gases CO_2 or CH_4. In a balanced ecosystem, NEP is equal to zero. If NPP is greater than the sum of nonplant respiration, NEP is positive, and the ecosystem gains carbon. Over the long term in most ecosystems, NEP is zero; otherwise organic matter would disappear (NEP < 0) or massively accumulate (NEP > 0), as it does in peatlands.

Supporting Ecosystem Models. The models of ecosystem energetics and carbon economy that support the use of chlorophyll *a* as an indicator are among the most mature and highly developed in ecology. This is one of the few areas in ecology where mechanistic predictive models already exist. Predictive phenomenological models of NPP were first developed in the 1960s and 1970s. For example, Leith's (1972) Miami Model showed that simple regressions of NPP against precipitation and temperature could predict most of the global variance of NPP. Phenomenological models continue to be important today because imagery collected for the entire Earth since 1972 by Landsat satellites permits the calculation of the Normalized Difference Vegetation Index (NDVI), which, as pointed out in Chapter 2, is a measure of chlorophyll per unit area. NPP is approximately proportional to NDVI, but a different proportionality constant is necessary for each vegetation type (Dai and Fung 1993).

Mechanistic models of NPP have also been developed by physiological ecologists during the past two decades. Farquhar et al. (1980) developed a model in which the rate of photosynthesis is the minimum of two different reactions—light-limited energy capture and carbon-limited CO_2 fixation. This model subsequently was extended and refined to include C_4 and C_3 photosynthesis (e.g., Collatz et al. 1992). It predicts the rate of carbon gain (moles/square meter of leaf area/unit time) as a function of leaf temperature, light level, and the internal concentration of CO_2. The model of Ball et al. (1986) predicts stomatal conductance as a function of net photosynthetic rates, CO_2 levels, and the humidity gradient across the plants' stomates. When coupled with simple physical equations governing the energy balance of leaves and the diffusion of gases through stomates, one can solve the equations directly for the photosynthetic rate and rate of transpiration as a function of temperature, humidity, and light level (see Foley et al. 1996, for a clear example).

To predict NPP, also needed are a model of plant respiration, a model of nutrient and soil moisture limitation, and a physical description or model of the vegetation (e.g., number of leaf layers and allocation). Models that predict NPP mechanistically are described by Sellers et al. (1997), Haxeltine and Prentice (1996), Raich et al. (1991), and Neilson (1995). To predict nutrient limitation, it is necessary to be able to predict decomposition of organic matter. The most general model of decomposition is the CENTURY model of Parton et al. (1988). Because mechanistic models of NPP predict decomposition, they necessarily also predict NEP.

Mechanistic models of NPP and NEP have been compared in a formal exercise (VEMAP 1995). They are as accurate as phenomenological models in current climates, but have the advantage that they predict NPP under novel conditions with some theoretical justification. However, when forced with novel climates, the models in the VEMAP comparison diverged from one another. This finding indicates that predictions for novel conditions need to be interpreted with caution and that the mechanistic understanding underpinning the models is incomplete in important ways. Methods of estimating NEP account well for above-ground carbon storage, but they do not account for storage of carbon in soils. Therefore, data on soil organic matter need to be added to the outputs of these models.

Numerical Characteristics of the Indicators. Chlorophyll per unit area ranges from 2.8 g/m^2 in tropical forests and 2.6 g/m^2 in temperate forests to 0.5 g/m^2 in tundra and desert ecosystems. Values of NPP in natural ecosystems range from 1,400 g/m^2/year in tropical rain forests (half that amount in temperate forests) to 50 g/m^2/year in American deserts (Whittaker and Likens 1973). NPP is markedly affected by land use, with values as high as 6,000 g/m^2/year in some agricultural systems and as low as zero in urban centers (VEMAP 1995). NEP for North America is thought to be currently 0.3×10^{15} g/year (IPCC 1995). It is not zero primarily because of the regrowth of forests cleared in the last century and early this century. Regrowing temperate forests typically have NEPs of 200-500 g/m^2/year (Wofsy et al. 1993). In contrast, tropical forests are thought to have a negative NEP currently because of deforestation (a total of roughly -1.0×10^{15} g/year [IPCC 1995]).

To obtain the aggregated national-level annual indicators of production capacity (total chlorophyll), NPP, and carbon storage, the values computed for each land cover category are summed. Annual changes in these numbers, rather than their absolute values, are the variables of interest because they reveal whether the productivity of the nation's ecosystems is being maintained and whether the total amount of carbon being stored is increasing or decreasing.

Temporal Variability. Three types of temporal variability must be taken into account in developing the NPP and carbon storage indicators:

1. **Daily Variation**. NPP and NEP vary daily because of variation in light and, to a much lesser extent, because of variation in temperature and

humidity. Obviously, NPP and NEP are negative at night and positive during the day. Chlorophyll density is stable diurnally.

2. **Seasonal Variation**. NPP, NEP, and chlorophyll density show dramatic seasonal cycles, with high values during summer and rainy seasons and low values during winter and dry seasons, when many plants shed their leaves. For calculating these indicators, yearly high values are the appropriate numbers.

3. **Interannual Variation**. All three indicators vary dramatically interannually. Natural disturbance or land use change obviously alters NEP, NPP, and chlorophyll density. For example, after a blow-down, a patch of forest has negative NEP for more than a decade, followed by a long period of positive NEP. NPP and chlorophyll density typically recover more rapidly than NEP (Schlesinger 1997). Even without disturbance or land use change, interannual variation in weather and climate can have dramatic impacts on these indicators. For example, Earth's terrestrial NEP currently fluctuates by at least 3×10^{15} g/year from one year to the next (Sarmiento et al. 1995), partly in response to El Niño events. NDVI measurements show that, over the past decade, the length of the growing season (period of high chlorophyll density) has increased in the North Temperate Zone by 10 days (Myneni et al. 1997).

Spatial Variability. Three types of spatial variability must also be considered in developing NPP and NEP:

1. **Small Scale (individual plant to stand)**. The most important cause of small-scale variation in NPP, NEP, and chlorophyll density is disturbance. An individual tree fall creates a patch of low chlorophyll density and negative NEP, which is subsequently converted to a patch of high chlorophyll density and positive NEP. Thus, stand-level indicators represent the average of a wide range of values at smaller scales.

2. **Medium Scale (stand to region)**. Within the same climatic region, the primary sources of spatial variation are changes in land use and variation in topography and soils. Each of these may cause stand-level variation in NEP, NPP, and chlorophyll density as large as those caused by disturbances or changes in climate.

3. **Large Scale (region to globe)**. Large-scale differences are driven by climate and associated land use and disturbance. Regional changes in chlorophyll, NPP, and NEP can be computed whenever it is desirable to do so, provided that the data are archived in a disaggregated form.

Data Requirements. NOAA and NASA satellites routinely acquire the imagery necessary to evaluate chlorophyll density at scales as small as 100 m^2 or less (see Chapter 2). Simple phenomenological models and the

CASA model (see below) can convert these images to estimates of NPP and NEP. In addition, four national programs are already in place that acquire the field data needed to estimate NPP and NEP by inventory methods.

The U.S. Forest Service's Forest Inventory and Analysis (FIA) program remeasures and censuses trees on more than 13,000 small plots (0.1 to 1 ha) every 10 years (see Chapter 2, Ground-Based Measurements). These data can be used to calculate forest NPP, but the locations of the plots and the data are difficult to obtain. If this program were augmented to include measures of soil carbon, it would yield direct estimates of forest NEP.

The Forest Service's Forest Health Monitoring Program collects information on forested ecosystem condition and production on a national network of 4,000 1-ha plots. The measurements again provide useful estimates of NPP, but because there are no measurements of belowground carbon, these data cannot provide direct estimates of NEP. However, the stand-structure data can be used to constrain predictive forest-gap models such as LINKAGES (Post and Pastor 1996) or the coming generation of physiologically grounded models (Hurtt et al. 1998) to provide estimates of NEP.

The USDA's National Resources Inventory provides a comprehensive assessment of the state and performance of natural and agricultural ecosystems on private lands every five years. Inventory measurements from 800,000 sites and regression models are used to estimate productivity, carbon storage, biomass, land use, vegetation cover, species ranges, and characteristics of soils.

Finally, the Long Term Ecological Research (LTER) network of 21 sites provides in-depth information on NPP, NEP, nutrient cycling, organic-matter dynamics, and disturbance (see Chapter 2). The spatial coverage of this network is obviously limited, however, representing only 19 sites in the United States and Puerto Rico.

In addition to sites providing direct inventories of carbon, the new Ameriflux network provides measurements of NEP and NPP using the new technique of eddy correlation. Each of the 24 sites in the Ameriflux network contains an eddy-correlation tower with a vertical array of sonic anemometers and CO_2 sensors that continuously measure the vertical gradient of CO_2 and the rate of vertical air flow. These data allow NEP to be computed from the amount of CO_2 that is removed from or added to the air by the ecosystem. One note of caution is that, for the two sites for which both eddy-correlation and inventory estimates of NEP have been calculated (Oak Ridge and Harvard Forest), the two methods differed by 50 to 100 percent (S. Wofsy, Harvard University, and W. Post, Oak Ridge National Laboratory, personal communications 1998).

At least two kinds of additional data are likely to significantly enhance our ability to estimate NPP and NEP in the near future. First, high-resolution satellites and the new canopy VCL satellite (see Chapter 2) will provide powerful additional data. Second, aircraft measurements of the vertical profiles of CO_2 in the atmosphere above an array of locations will allow NEP to be estimated by tracer-transport inversion. This technology represents a large-scale analog of the method used in eddy-correlation towers. If wind speed and direction across the parcel are known from weather data, and the amount of fossil fuel consumed in the region is also known, then the CO_2 gradient in the atmosphere over the parcel can be estimated, assuming a balanced ecosystem (NEP = 0). If the actual gradient is less steep than the expected gradient, then the ecosystem within the parcel is consuming CO_2, and NEP is positive. Tracer-transport inversion currently allows resolution only at continental scales, but planned arrays of measurements will allow tracer-transport inversion estimates of NEP for separate regions within the United States.

Two other groups of models can provide useful estimates of ecosystem performance. First, gap models, which were first developed in the 1980s (Shugart 1984 and the references it cites), are now being constructed with the same mechanistic underpinnings as the NPP-NEP models (Hurtt et al. 1998). These new models should predict population dynamics with the accuracy of gap models, and ecosystem functioning with the accuracy of mechanistic NPP-NEP models. Additionally, Potter et al. (1993) have developed a model that is intermediate between mechanistic and satellite models of NPP-NEP. This model, called CASA after the participating institutions (Carnegie, Stanford, and NASA-Ames), uses satellite NDVI to constrain estimates of production. However, it also includes mechanistic models of respiration, allocation, and decomposition that allow the computation of NEP. Although the CASA model cannot predict the future (because it requires temporal sequences of satellite imagery as a driving variable), its mix of mechanism and reliance on data make it ideal for relating chlorophyll density, NPP, and NEP.

An Indicator of Aquatic Productivity—Trophic Status of Lakes

People tend to settle around water and to discharge wastes, treated or not, into lakes, rivers, bays, and estuaries. As a result, the structure and functioning of aquatic ecosystems have been highly modified in most parts of the United States. The status of some of the nation's lakes is currently monitored by EPA and other agencies, but a suitable national indicator of lake status has not yet been devised. Such an indicator can be developed from the fact that a few fundamental characteristics determine the functional properties of lakes and their ability to provide the many goods and services valued by society. The key characteristics are themselves closely interrelated: nutrient status, rate of biological production, and net biological production. Together, these characteristics define a lake's trophic state.

Together with basic physical conditions, such as morphology and hydrology, primary production in lakes is determined by inputs of energy and chemicals—most importantly nutrients, inorganic ions, and natural organic matter. Lakes generally act as traps or sinks for substances, some of which are exported, but many of which are deposited in sediments. These sediments serve as repositories of information on the history of watershed conditions and the ways a lake has responded to inputs from its watershed.

Lakes that have low concentrations of nutrients, low rates of primary production, and generally low standing crops of plants, are called *oligotrophic*. Conversely, lakes with high nutrient levels, high plant production rates, and an abundance of plant life are called *eutrophic*. Lakes intermediate in these characteristic are called *mesotrophic*, and extremes on the ends of the trophic state continuum are termed *ultra-oligotrophic* and *hyper-eutrophic* respectively.

The most appropriate trophic state condition of a lake varies according to the goods and services expected of it. For drinking water, the more oligotrophic a lake is, the better, because such waters are easier and less expensive to treat. Recreational values also decline as lakes become more eutrophic, because nuisance algal blooms, odor problems, and fish kills occur with increasing frequency, and water clarity decreases to the extent that swimmers no longer find the lake attractive. On the other hand, higher concentrations of nutrients, and thus higher rates of primary production, generally yield higher rates of fish production. However, as primary productivity increases, the nature of the fish community changes, and eventually less desirable fish species become dominant. The optimal trophic state for fishing thus depends on the type of fish desired.

Lakes across the nation suffer from many stresses caused by human activities. Atmospheric deposition of acidity affects many softwater lakes in the Northeast and Upper Midwest. High concentrations of mercury in fish, also derived from atmospheric deposition, are also widespread. Suspended sediment from soil erosion causes poor water clarity in many reservoirs of the Southeast and Great Plains states. However, the most widespread, human-induced problem for lakes in the United States is excessive nutrient enrichment or eutrophication: it accounted for 43 percent of the 6.7 million acres of lakes found to be impaired in the most recent survey of national water quality (U.S. EPA 1998). Eutrophication affects lakes of all types and sizes and in all geographic regions, including backwater areas and impoundments on many rivers and in coastal and estuarine waters. It is closely associated with intensive human disturbance of watersheds through urban development and agriculture.

Of the many biological, chemical, and physical characteristics of lakes that vary along the trophic continuum, three are commonly used as indicators of the trophic status of lakes: Secchi-disk transparency, total phosphorus, and chlorophyll *a*. These indicators respectively measure physical, chemical, and biological characteristics of lake status. Secchi-disk transparency measures water clarity in terms of the depth at which a white disk (20 cm in diameter) is just visible in the water column. In oligotrophic lakes, Secchi-disk transparency generally is greater than 3 meters; in ultra-oligotrophic lakes, such as Lake Tahoe, California, transparency values can exceed 20 meters. Transparency values during summer generally are less than 2 to 3 meters in eutrophic lakes and less than 1 meter in hyper-eutrophic lakes. In extreme cases, transparency is reduced to a few centimeters. Although Secchi-disk transparency is a technologically crude measurement, it has great appeal as an indicator of lake water quality because it measures a condition directly related to human perception of lake quality. Furthermore, because decreased transparency in most lakes is caused by algal growth, transparency is also a good measure of lake status.

Phosphorus generally is the most important nutrient limiting plant growth in lakes, and total phosphorus (TP) is the most commonly used measure of trophic state. Oligotrophic lakes generally have TP concentrations of less than 10 μg/L, and eutrophic lakes TP concentrations greater than 20 to 30 μg/L (the criterion separating trophic state classes varies somewhat by geographic region). Hyper-eutrophic lakes may have TP concentrations as high as several hundred μg/L.

The concentration of chlorophyll *a* is a direct quantitative measure of algal densities in a lake. Average chlorophyll *a* concentrations during the growing season in oligotrophic lakes, which have low phytoplankton densities, usually are less than 5 μg/L. Surface waters in eutrophic lakes have average summer chlorophyll *a* concentrations greater than about 10 μg/L; peak concentrations during algal blooms may exceed 50 μg/L.

Values of the three indicators vary seasonally in lakes. Although many limnologists argue that at least 5-6 measurements are needed over the growing season to accurately assess the trophic state of a given lake (e.g. Heiskary and Walker 1988; Brown et al. 1998), less-frequent sampling is sufficient for large regional surveys. One set of measurements during an index period should be adequate if measurements are done consistently from year to year on a well-defined (and large) sample of lakes. The critical period of recreational use for north temperate lakes is approximately the beginning of July to the end of August. Serendipitously, the variance in trophic state variables appears to be lowest during this period, and maximum chlorophyll *a* levels and minimum Secchi-disk transparency values typically occur during this period (Kloiber et al. in press).

Moderate interannual variations occur in the indicators in response to variations in climatic and hydrologic of factors. Response times for the indicators to changes in external nutrient loadings generally is rapid (few years or less) provided that the maintenance of eutrophic conditions in a lake is not controlled by internal recycling of phosphorus from nutrient-rich bottom sediments. Consequently, in many cases it is possible to detect a trend in the trophic status of a given lake from annual data collected over only a few years.

Current Data Collection Efforts. Because Secchi transparency is easy and inexpensive to measure much information is available on lakes, and many states have citizen monitoring programs that rely primarily on transparency measurements. Abundant data also are available on chlorophyll *a* and TP, but there currently is no organized program to obtain, store, and analyze lake trophic-state data on a national basis. Instead, data are gathered by a wide range of governmental units, primarily state natural resource and environmental protection agencies, municipalities, park boards, and watershed districts, as well as private lake associations, consulting firms, and university researchers.

The Clean Water Act (Section 305b) requires states to report every two years on the quality of their lakes to the U.S. EPA, which in turn submits a report to Congress (e.g., U.S. EPA 1998). However, there are no prescribed standards for data collection and reporting, and the sequence of reports cannot be used to gauge temporal trends because sampling units and reporting methods vary inconsistently over time. Indeed, it is questionable whether the 305b reports represent an accurate assessment even of lake status at a given time because sampling programs generally are not designed to allow data to be extrapolated to the population of lakes in a state as a whole.

The three primary measures of lake status are strongly correlated: TP and chlorophyll *a* are related in a positive log-linear fashion and Secchi-disk transparency is related in a negative hyperbolic fashion to chlorophyll *a*. The three indicators have been transformed into simple indices that express one concept of trophic state quantitatively (Carlson 1977). Some states use these indices to classify their lakes according to trophic state; for example, Minnesota uses the following Trophic State Index (TSI) ranges to categorize lakes: less than 40, oligotrophic; 40-50, mesotrophic; 50-70, eutrophic; and more than 70, hyper-eutrophic (Heiskary and Walker 1988). There has been considerable development and use of the TSI (Carlson and Simpson 1996) and related methods for collecting data. Where possible, we recommend calculating the TSI from measurements of all three factors: TP, Secchi depth, and chlorophyll *a*. TSI should include observed values of chlorophyll *a* where possible, since chlorophyll *a* provides the most direct measure of biological activity. Deviations from the expected relationships among chlorophyll *a*, total phosphorus, and Secchi-depth transparency signal regional variations in water color, factors other than primary production that limit transparency, or limitation of production by some nutrient other than phosphorus.

TSI can be aggregated nationally by computing a frequency distribution of trophic states across lakes. The frequency distribution of trophic states (but not an average of TSI values) should be used as a national-level indicator, because changes in this distribution provide the most useful information. The number of lakes that become more eutrophic over time would clearly indicate mismanagement of fertilizers, sewage, and other sources of nutrients. A national increase in numbers of oligotrophic lakes would be an indication of better management of nutrients. Threshold values can be used as an additional way to direct attention to important trends in lake status. For example, increases in trophic state are not always undesirable, but hyper-eutrophic lakes have fewer valuable properties than other lakes. Therefore, the national indicator should also record the number of lakes that are hyper-eutrophic, in addition to reporting changes in frequency distributions of lakes and trophic states.

Example of the Use of the Indicator. Volunteer monitoring programs are well established in many states. These programs have provided a useful baseline to assess the biological condition of lakes. TSI has been incorporated into many volunteer lake-monitoring programs, because of the low cost and simplicity of collecting Secchi-disk transparency readings. Although Secchi-disk transparency does not provide a complete measure of trophic state, it provides a repeatable estimate of trophic state across a range of conditions. Some volunteer monitoring programs include chlorophyll *a* and total phosphorus measurements to the array of samples collected by volunteers.

The distribution of lakes' trophic states should be computed using an unbiased sample of lakes. However, lakes assessed by volunteers are generally biased toward recreationally accessible lakes. Agencies and municipalities generally assess lakes that provide an important local or regional resource such as drinking water. Conservation organizations and park services generally have an interest in assessing more pristine waters. Therefore, careful attention needs to be paid to the statistically reliable selection of lakes to be represented in the national index of lake status, as has been done in the selection of lakes for regional studies of the effects of acid deposition (Linthurst et. al. 1986).

Information from the large number of lakes already sampled can be assembled to develop a baseline frequency distribution of lake trophic state. Because the Great Lakes contain so much more of the fresh water in the United States than all other lakes together, the data on these lakes should be computed and reported separately, using the many stations that are already used to record conditions in different portions of the basin under different land influences. The trophic state of each of the Great Lakes should be presented as the distribution of values among that lake's sampling stations as well as their average value, because changes in TSI values at those stations indicate changes in management of portions of each of the Great Lakes.

TSI can provide sufficient resolution to detect changes in average trophic state of a few percent when as few as 50 lakes are sampled (NRC 1994). However, given the low cost and skill required to collect these data, it is feasible to monitor thousands of lakes nationwide. A nationwide collection of TSI values on several thousand lakes on an annual basis would provide an accurate and precise measure of trends over periods of less than 10 years. The distribution of trophic states needs to be reported and interpreted in two ways: as overall number of lakes in various trophic states, and as numbers of lakes that have changed and the direction of that change. The distribution of trophic states of lakes is a useful national indicator, but the distribution of trophic states can also be compiled regionally or locally relatively simply.

How Will Technology Developments Affect This Indicator? The use of remotely sensed data could provide some measurements to replace in situ sampling. Remote sensing is likely to be most useful for measuring inaccessible lakes and providing quality control for volunteer monitoring efforts. Chlorophyll *a* extraction methods have recently been simplified so that field filtration and extraction can be done with only minimal training. Samples would still need to be measured spectrophotometrically or fluorometrically in the laboratory. In situ fluorometry is technologically possible, but it provides a more qualitative than quantitative estimate of chlorophyll *a*. In situ probes with fluorometry probes would be useful to measure within-lake variance in chlorophyll, but would require confirming calibration samples to provide quantitative between-lake comparisons.

Previous studies (e.g., Brown et al. 1977a, and b; Lillesand et al. 1983) have found good correlations between satellite reflectance data and water clarity and chlorophyll *a* concentrations. Recent improvements in satellite technology and especially in the software to process images quickly and efficiently, coupled with much lower costs for the images themselves, indicate great potential for use of satellite imagery to gather trophic state information on a broad regional or nationwide basis (Kloiber et al. in press). Satellite data cannot replace ground-based (in-lake) measurements altogether; there is a continuing need for the latter information to calibrate reflectance data from satellites. However, satellite imagery has the potential for extending trophic state assessments virtually to every lake in a region in a very cost-effective manner.

Both Secchi-disk depth and phosphorus measurements are well developed and standardized. They are unlikely to be replaced by remote sensing or other field techniques. Various electronic devices can measure turbidity and transparency, but not more cost-effectively than a Secchi disk used by a volunteer.

Improvements to personal computers and the Internet will enhance the effectiveness of the collection of data by a combination of government and volunteer monitors. Advances in personal computing will allow data collection and functions to be standardized and stored in a central database that can receive data from a large number of volunteers. Software can also standardize quality control and data verification.

An Indicator of Trophic Status of Streams—Stream Oxygen

Indicators of the status of streams could be based on models of flowing-water ecosystems, such as the River Continuum Concept (RCC), which is a broad, integrative framework for conceptualizing stream-riparian systems (see Chapter 2). However, the ability of the RCC to explain or predict primary production and respiration is limited, because of the complex relationships among terrestrial production, river size and flow rate, and inputs to instream processes. An indicator that better captures the balance between instream primary production and respiration is dissolved oxygen (DO) concentration, or *stream oxygen*, whose values can be directly related to NPP and respiration (R). If NPP is much greater than R, excessive plant growth (algal blooms, stream channels clogged by macrophytes) usually results; if NPP is much less than R, the result is low DO concentrations, often followed by fish kills and undesirable odors.

The concentration of DO in a river at any time is a complicated function of temperature and various sources and sinks for oxygen. Oxygen solubility ($S[O_2]$) in water decreases in a nonlinear fashion with increasing temperature (thus, $S[O_2]$) = 14.5 mg/L at 0°C and 9.2 mg/L at 20°C). The main sources of oxygen in water are photosynthesis and atmospheric re-aeration (which occurs only when the oxygen concentration in the water is less than its solubility at a given temperature and pressure). The main sinks of oxygen are microbial respiration, nitrification (bacterial oxidation of ammonium to nitrate), plant respiration, sediment oxygen demand, and degassing to the atmosphere (when the water is saturated with oxygen). Ambient concentrations much greater than the saturation value at a given temperature and pressure indicate a preponderance of photosynthetic activity and the likelihood of high nutrient concentrations, algal blooms, and/or excessive growths of macrophytes. Ambient concentrations much below the saturation value indicate a preponderance of respiration and the likelihood of organic enrichment from wastewater or from a high rate of plant production upstream caused by high nutrient levels. Many states set minimum standards for DO at values that generally protect fish and other aquatic life. Typical minimum standards are around 5 mg/L, but in streams with particularly sensitive species such as trout, higher minimum standards are used.

Spatial and Temporal Variability. Because stream oxygen can change rapidly in surface waters, care must be taken to obtain data that can be interpreted and compared with data from other sites. Because oxygen concentrations vary significantly during the course of a 24 hour period, monitoring should be done at the same time of day or, preferably, continually over a 24 hour period. To aid in the interpretation of the oxygen data, data collection should include a few other basic water chemistry variables. Electrical conductivity (a measure of the total ionic content of the water), pH, and turbidity, which are all easy to measure, constitute the minimum set of chemical measurements that should accompany measurements of oxygen in a national stream–monitoring program.

Current Data-Collection Efforts. Most states monitor DO as part of ambient monitoring programs. Additionally, federal agencies, such as the Environmental Protection Agency, the Army Corps of Engineers, and the Geological Survey, monitor DO at selected sites. Furthermore, permitted dischargers sometimes monitor DO in receiving streams.

Example of the Use of the Indicator. Stream oxygen can be used in two ways to indicate the state of flowing-water ecosystems. DO is an important environmental requirement of fish and other aquatic life. Where DO concentrations are low, 3 to 4 mg/L or less, fish reproduction can be severely affected, and invertebrates that process organic matter can decline or be extirpated. Very high DO concentrations (supersaturation) can result from excess oxygen production either as a result of eutrophication or as a result of elevation of total dissolved gases. Elevation of total dissolved gases has been associated with gas-bubble trauma and increased mortality of fishes below large dams.

Oregon bases part of its water-quality index on the unweighted harmonic mean square of various subindexes, including one for stream DO (Cude 1996, Doljido et al. 1994). The stream DO subindex is based on the oxygen concentration for measurements that are at or below 100 percent saturation and on the percent saturation if oxygen measures are above 100 percent saturation. This index value is scaled from 0 to 100 and is monotonic and positive for concentrations from 3.3 to 10 mg/L. For supersaturation, the index value decreases as saturation increases from 100 to 275 percent.

How Will New Technology Affect this Indicator? One recent technological development that is likely to affect the use of stream oxygen measurements is the increasing availability of reliable probes and devices for telemetering data, which allow continuous monitoring, data storage, and remote access to data. As the cost of these technologies drops, the cost of data acquisition will decline, so that the principal obstacle will be the time needed to change the devices' batteries. Improved technology should allow more sophisticated and varied use of stream oxygen as an indicator. For example, natural patterns of primary production and respiration can produce day-night differences in stream oxygen. The absence of such differences should be cause for concern during much of the year. The new technologies described above should make the detection of those differences or their absence much easier for most sites by reducing the current need for extraordinary sampling effort.

INDICATORS OF NUTRIENT-USE EFFICIENCY AND NUTRIENT BALANCES IN AGROECOSYSTEMS

An agroecosystem is an ecosystem managed intensively for the production of food or fiber. Cropland, pasture, and range make up more than 55 percent of the total land area of the contiguous United States (USDA 1997). Agroecosystems are typically studied at field, whole-farm, and regional scales (FAO 1994); methods being developed in precision agriculture require studies at the subfield scale. In developed countries, agroecosystems are managed for high production and are characterized by intensive nutrient and pesticide inputs, fast growth-harvest cycles, and low genetic diversity of crops and animals (Odum 1984, Matson et al. 1997). These intensive agricultural management practices can have adverse impacts on soil quality, and they generate offsite effects on surface water, groundwater, and the atmosphere. Agroecosystems in the United States also contain parcels of unmanaged or lightly managed areas such as woodlots, fencerows, and riparian areas.

The soils of agroecosystems have long been the focus of national conservation efforts to reduce losses by erosion. More recently, concerns have expanded beyond erosion to include salinization, compaction, loss of soil organic matter and attendant desirable physical properties, and the accumulation of trace elements and toxic substances (NRC 1993). Over the past five decades, increased demand, trends in dietary preferences, and the development of new technologies have led to greatly expanded use of chemical fertilizers, pesticides, and water for irrigation and livestock production. Research to find ways to reduce the deleterious effects of these practices on surface and groundwater quality and on soils is leading to policies to protect soil and water while sustaining the profitable production of agricultural goods, but considerable improvements can still be made (NRC 1993).

Nutrient cycles differ dramatically between agricultural ecosystems and the natural ecosystems they replaced. The most significant recent change involves massive movements of nutrients across landscapes. For example, fertilizers are transported to crop-producing areas in the spring, grain is transported to animal-producing areas in the fall, and animal manures become wastes or excess fertilizer because the rate of production exceeds local needs and the cost of transport make redistribution economically infeasible (Magdoff et al. 1997).

More land, fertilizer, pesticides, and irrigation water are needed to support animal production, and the environmental impacts are greater than if dietary choices demanded less animal protein. The importance of animal protein in human diets, which is consumer-driven, is an important factor in agriculture's impacts on the biosphere. Nations, such as the United States, with extensive concentrated animal production facilities generate large amounts of excess nutrients because nutrient use in animal production is much less efficient than in producing crops (van der Ploeg et al. 1997).

The NRC (1993) analyzed agricultural practices and impacts to identify opportunities that held the most promise for "improving the environmental performance of farming systems while maintaining profitability." The broad recommendations of that report were the following:

1. Conserve and enhance soil quality as a fundamental first step to environmental improvement.
2. Increase nutrient, pesticide, and irrigation efficiencies in farming systems.
3. Increase the resistance of farming systems to erosion and runoff.
4. Make greater use of field and landscape buffer zones.

Following this framework, the committee evaluated and recommends national-level indicators of *nutrient-use efficiency* and *balance*.

The high productivity of most modern agriculture depends on added nutrients. In most agroecosystems, more nutrients are added to the system than are extracted from it in harvested products; and the imbalance is far greater for animal production than for crop production. These excess nutrients find their way into the soil, the atmosphere, and water.

We define the proportion of added nutrients removed in products as *nutrient-use efficiency*. Because the efficiency with which nutrients—especially N and P—are used in the production of crops and animal products is of great economic and environmental significance, it is important to monitor changes in inputs and outputs from agricultural lands. Losses of agricultural chemicals account for a major share of nonpoint-source N and P pollution of ground and surface waters (NRC 1993). Because point-source control of N and P inputs to surface and groundwaters has been easier to achieve, nonpoint sources account for an increasing share of the total inputs (Sharpley and Meyer 1994). The increasing demand for agricultural products will generate powerful pressures for increased agricultural chemical use.

N export from agroecosystems is known to adversely affect drinking-water supplies. Nitrate (NO_3) in drinking water can be acutely toxic, and it can cause methemoglobinemia in infants (Spaulding and Exner 1993). The maximum contaminant level has been set at 10 mg NO_3–N L^{-1} for drinking water. Regions of irrigated agriculture such as the wheat belt and California's Central Valley have the highest incidence of elevated NO_3 levels, but many other such areas are scattered about the United States (Spaulding and Exner 1993, Kolpin 1997, Lichtenberg and Shapiro 1997). N also contributes to eutrophication of aquatic and estuarine systems (Spaulding and Exner 1993).

Local, regional, and national annual N budgets for cropland and agricultural watersheds indicate that more N is added to croplands (manure N plus fertilizer N plus N fixed by legumes) than is removed in crops. The amount of N not transferred to crops varies widely as a function of site conditions and management practices. Site conditions that control N transformations vary among regions, along topographic gradients, and with different cropping practices. Also gaseous losses from and additions to the soil are difficult to measure; and cropping practices other than fertilization can release stored soil N at substantial rates (Keeney and DeLuca 1993, David et al. 1997). Therefore it is difficult to rigorously determine the fate of excess fertilizer and manure N, and difficult to relate the excesses directly to elevated ground- and surface-water N levels (Keeney and DeLuca 1993, David et al. 1997, Kolpin 1997).

Gaseous losses of N (as N_2O and NH_3) have other influences. N_2O is a significant greenhouse gas. NH_3, volatilized from fertilizers and animal manure, is eventually deposited as NH_4. After being taken up by plants, atmospherically deposited NH_4 acidifies the soil and, in some regions (e.g., the montane watersheds of the northeastern United States), contributes substantially to the N and H^+ loads that must be assimilated by sensitive landscapes (Johnson and Lindberg 1992, Vitousek et al. 1997).

Soil phosphorus lost from agricultural systems to watersheds accelerates eutrophication of lakes and streams. Productivity and algal blooms in lakes and streams are promoted by elevated inputs of dissolved inorganic P, and by labile P ("algal-available P") bound to sediments or in labile organic combinations. In general, P is less mobile than N, because it adheres strongly to soil constituents. P applied in excess of crop uptake is retained to a considerable extent in nonsandy mineral soils. Quantitatively important leaching losses of P through the soil to ground and surface waters are limited to areas of sandy soils, organic soils, and cases of extreme P loading. Surface runoff and tile-drain effluent from fertilized cropland, pastures, and animal-feeding operations and the accompanying suspended sediment are thus the most important vectors for P delivery to surface waters. The importance of each source varies according to region, soil conditions, and management practices.

As concern about environmental impacts of agricultural nutrient losses has grown over the past decade, so has research on how to manage agricultural nutrient use to minimize loss, while maintaining productivity and profitability. Because nutrient loss from agricultural systems is a site-specific problem (NRC 1993, Harris et al. 1995), site- and practice-specific mitigation measures are required. Many efforts are under way to gain the understanding necessary to balance environmental and productivity needs (see reviews by NRC 1993, Harris et al. 1995, Daniel et al. 1998, Sharpley et al. 1996).

Changes in nutrient-use efficiency in agricultural systems and nutrient losses from these systems have been driven by the increased availability of chemical fertilizer since World War II; by the availability of transportation for fertilizer, feed, and agricultural products; and by increases in meat consumption, which has led to the growth of specialized animal-production systems. The trend toward greater livestock production in the Western world during the latter half of this century is a major contributor to overall loss rates of nutrients used in agriculture, because the nutrients in manure are much less efficiently incorporated into animal products than into crops (e.g., van der Ploeg et al. 1997). This situation is magnified in small meat-producing countries such as the Netherlands, where animal feeds grown on five to seven times the Dutch agricultural land area are imported. This large import of nutrients is driving country-wide nutrient enrichment as the manure is applied to Dutch agricultural land and excess nutrients make their way into ground and surface waters (Van der Molen et al. 1998).

The Indicators. Nutrient leakage is an inherent property of current agricultural activities (see review by Magdoff et al. 1997) and will remain so for the foreseeable future. Because the demand for agricultural products will increase as the human population and economic activity increase, the only way to reduce losses of nutrients to ground and surface waters (and to the atmosphere in the case of gaseous N losses) is to develop and implement site-specific and practice-specific management techniques that improve the efficiency of nutrient use in crop-producing areas, and that limit the leaching and runoff of nutrients from animal-producing operations. Considerable agricultural research is being conducted on this matter; NRC (1993) covers useful management alternatives. If animal products become less important in people's diets, overall agricultural nutrient losses will decrease (van der Ploeg et al. 1997). However, even if substantial improvements in nutrient-use efficiency occur, overall losses from agricultural lands will increase if demands for agricultural products outpace improvements in nutrient-use efficiency. Accordingly, it is useful to have indicators of both the overall efficiency of nutrient use in the production of crops and animal products and the overall nutrient balance. Efficiency indicators are ratios or percentages that can increase or decrease with time. Balance indicators record the excess of nutrients applied to agricultural land over nutrients removed in harvested products. Indicators of nutrient-use efficiency and overall balance can be created for use at virtually any scale from farm fields to countries.

Nutrient-Use Efficiency. N and P use-efficiency indicators are useful for crops or industries, for counties, and for watersheds in which ground or surface waters are perceived to be adversely affected. These indicators can be used in testing trends in the effectiveness of management programs locally, regionally, nationally, or on a crop-specific basis. Aggregated data at a national scale have been useful for detecting and understanding trends in N-use efficiency in Germany (van der Ploeg et al. 1997). N and P budgets for cropland are often very hard to construct given the difficulties in tracking nutrients applied in excess of crop uptake, and especially in determining gaseous inputs and outputs of N. Different authors have used different indicators to represent fertilizer-use efficiency, and several assumptions are usually made in estimating N budgets for croplands (e.g., Jenkinson and Smith 1988, Black 1993, NRC 1993). We have adapted the approach and budget methods used by the NRC (1993) and van der Ploeg et al. (1997) in constructing the following indicators to monitor trends in agricultural nutrient-use efficiency at a national scale.

For cropland:

(1) Nitrogen-use efficiency Nc =

$$\frac{\text{N removed in crop biomass (mass y}^{-1}\text{)}}{\text{chemical fertilizer N applied + animal manure N applied + N fixed by legumes (mass y}^{-1}\text{)}}$$

(2) Phosphorus-use efficiency Pc =

$$\frac{\text{P removed in crop biomass (mass y}^{-1}\text{)}}{\text{chemical fertilizer P applied + animal manure P applied (mass y}^{-1}\text{)}}$$

Data inputs for N and P removed in crops are crop yields by type, dry-matter percentages, and biomass N and P content. Fertilizer sales by county (e.g., EPA 1990, Smith et al. 1997), county animal censuses (e.g., U.S. Bureau of Census 1987), and per-animal nutrient excretion rates (U.S. Soil Conservation Service 1992) can be used as estimates of the terms in the equations. Legume N fixation requires assumptions, but they are straightforward and useful if they are uniformly applied (NRC 1993). Because the indicators are percentages, they take values ranging from 0 to 100 percent, higher values indicating greater efficiency.

Depending on availability of data, these indicators could be calculated annually and should be able to detect changes in nutrient-use efficiency on a decadal scale. For example, in the former West Germany, crop N-use efficiency, calculated as shown above (without N fixed by legumes), decreased from about 100 percent in 1964 to a low of 72 percent in 1984, then increased to about 81 percent in 1990 (van der Ploeg et al. 1997). Using NRC (1993) data for 1987 (NRC's Table 6-3, medium N-fixation scenario), N-use efficiency for U.S. cropland was 59 percent and for P 31 percent. Because such large quantities of nutrients in excess of those harvested are applied to U.S. agroecosystems, the potential for improvements in nutrient-use efficiency is great (NRC 1993).

Nutrient-use efficiency indicators should provide integrated measures of how well management strategies are working. Application of the indicators at smaller scales would provide opportunities to target specific geographic areas or crop-production systems. A rough but potentially useful measure of the efficiency of N and P use in agriculture on a national scale can be approximated by the equation

(3) Na =

$$\frac{\text{N content of crops produced for human consumption + N content of animal products produced (mass N y}^{-1}\text{)}}{\text{Chemical N fertilizer applied to cropland + N fixed by legumes (mass N y}^{-1}\text{)}}$$

$$(4)\ \mathrm{Pa} = \frac{\text{P content of crops produced for human consumption} + \text{P content of animal products produced (mass P y}^{-1}\text{)}}{\text{Chemical P fertilizer applied to cropland (mass P y}^{-1}\text{)}}$$

These indicators could be applied to smaller scales such as states, counties, or watersheds, but at these smaller scales the indicators would need to be modified to account for imported (or exported) nutrients in crops produced for animal feed. The large decrease in N-use efficiency for agriculture as a whole (i.e., animal and crop production) from 1951 (77 percent) to 1980 (27.2 percent) in West Germany resulted from increasing rates of applications of chemical fertilizers, the rise of livestock production (a much less efficient user of nutrients than crop production), and an attendant 10-fold increase in imported N in feed (van der Ploeg et al. 1997).

Nutrient Balance. The same data can be used to compute indicators of overall nutrient balance. Changes in these indicators reflect the combined effects of changes in nutrient-use efficiency and in the quantity of agricultural products. Tables 4.3 and 4.4 show the mass-balance approach used by the NRC (1993) for N and P applied to cropland. Such budgets require several assumptions, but they are straightforward. For nitrogen, the indicator would be N fertilizer applied plus recoverable N in animal waste (manure) applied plus N fixed by legumes minus N removed in harvested crops (assuming steady-state nutrient pools in crop residues). A similar indicator could be used for P, as described for the nutrient-use efficiency indicator. These national-level indicators would provide important input for understanding changes in water quality and other aspects of nutrient cycling. The units of the indicators are Mg/year for the area of interest.

Nutrient-balance indicators could also be used at local and regional scales. States and counties conduct censuses of farm animals, and methods are available for estimating per capita manure production by poultry and livestock (NRC 1993). The indicator would not be affected by variations in the distances between the animal production facilities and the croplands that support them. The indicator would have smaller values in areas of crop-only production and higher values where animal production is concentrated. Such indicators would help states and counties evaluate their agricultural practices in light of potential nutrient losses to surface water and groundwater.

TABLE 4.3 Regional and National Estimates of Nitrogen Inputs, Outputs, and Balances on Croplands, Medium Legume-N Fixation Scenario (metric tons)

	Input					Output			
Region	Nitrogen Fertilizer	Recoverable Manure-N	Legume-N	Crop Residues	Total	Harvested Crop	Crop Residues	Total	Balance
Northeast	252,000	224,000	284,000	70,200	831,000	412,000	70,200	482,000	349,000
Appalachia	564,000	109,000	395,000	102,000	1,170,000	491,000	102,000	593,000	577,000
Southeast	609,000	78,100	173,000	43,800	904,000	236,000	43,800	280,000	624,000
Lake States	988,000	278,000	984,000	368,000	2,620,000	1,420,000	368,000	1,780,000	834,000
Corn Belt	2,720,000	240,000	2,750,000	1,220,000	6,940,000	3,860,000	1,220,000	5,080,000	1,850,000
Delta	468,000	59,400	552,000	105,000	1,180,000	444,000	105,000	548,000	636,000
Northern Plains	1,510,000	256,000	1,020,000	602,000	3,390,000	1,930,000	602,000	2,530,000	859,000
Southern Plains	920,000	183,000	82,000	129,000	1,310,000	503,000	129,000	632,000	682,000
Mountain	554,000	160,000	451,000	158,000	1,320,000	786,000	158,000	944,000	379,000
Pacific	789,000	146,000	177,000	88,700	1,210,000	498,000	88,800	587,000	623,000
United States	9,390,000	1,730,000	6,870,000	2,890,000	20,900,000	10,600,000	2,890,000	13,500,000	7,420,000

Source: NRC 1993.

TABLE 4.4 Regional and National Estimates of Phosphorus Inputs, Outputs, and Balances on Croplands, 1987 (metric tons)

	Input				Output			
Region	Fertilizer-P	Recoverable Manure-P	Crop Residues	Total	Harvested Crop	Crop Residues	Total	Balance
Northeast	155,000	76,700	6,660	239,000	48,300	6,600	55,000	184,000
Appalachia	334,000	42,300	9,130	385,000	60,500	9,130	69,600	315,000
Southeast	264,000	32,400	3,960	300,000	27,800	3,960	31,800	269,000
Lake States	442,000	98,500	34,100	575,000	168,000	34,100	202,000	373,000
Corn Belt	1,130,000	115,000	112,000	1,350,000	481,000	112,000	593,000	761,000
Delta	119,000	22,200	10,600	151,000	54,600	10,600	65,200	86,200
Northern Plains	446,000	100,000	58,800	605,000	261,000	58,800	319,000	286,000
Southern Plains	274,000	59,900	14,300	348,000	72,400	14,300	86,700	262,000
Mountain	186,000	53,300	14,500	254,000	93,500	14,500	108,000	146,000
Pacific	224,000	54,300	8,450	287,000	57,700	8,450	66,200	221,000
United States	3,570,000	655,000	272,000	4,500,000	1,320,000	272,000	1,600,000	2,900,000

Source: NRC 1993.

RESEARCH NEEDS

Although the committee's recommended indicators are based on solid theoretical justification and extensive data, the precision and interpretation of the indicators should be improved by additional research and development. Moreover, future research may suggest new indicators that may be better than existing indicators or may be usefully added to the set of regularly reported indicators. More research is needed to identify organisms and biological processes that are especially sensitive to stresses and perturbations and to determine more accurately the temporal behavior of indicators. To address problems associated with spatial scale in indicator performance, it is important to test new indicators carefully in pilot programs before implementing them nationally (NRC 1995a). For the recommended indicators that are new, especially the ones that measure land cover, land use, and species diversity, further work is needed on how best to operationalize the indicators and to help identify future research plans. This work, which should include one or more workshops, should include academic scientists, practitioners, agency scientists, and other interested parties.

Temporal Behavior of Ecological Indicators

Knowing temporal variations in indicator values is important for interpreting monitored data. Because very few monitoring efforts exceed a decade, and many for no more than a few years, other methods need to be used to provide a record long enough to capture normal variations in the proposed indicator values, as well as to investigate surprise variations.

The source of much of the variation in indicators may be found at the interface between population and ecosystem processes. Population cycles are well studied, but how population cycles affect temporal patterns of species diversity, carbon storage and flow, and nutrient runoff is unknown. Experimental and theoretical investigations into the relationship between current population cycles and related ecosystem processes will provide the mechanistic understanding needed to improve predictions and interpret the causes of indicator behavior.

In addition, research is needed on applications of the mathematical techniques of spectral analysis (Platt and Denman 1975) and wavelet analysis (Wickerhauser 1994) to time series of model outputs and to patterns in the paleorecord. Spectral analysis identifies the periods and amplitudes of fluctuations in a time series of data. An example of how this method can assist in the design of sampling systems is provided in Appendix A. Wavelet analysis identifies periods in which there is a sudden change in system behavior. These periods may correspond to "surprises." These mathematical techniques have been developed for analysis of digital signals and other relatively clean data sets free from stochastic effects. Ecological data, in contrast, are often incomplete because of gaps in the record. In addition, in ecological research, stochastic noise often obscures the detection of fluctuations of long periods or low amplitudes, and the detection of sudden changes in the data spectrum. Research is required on adapting these mathematical techniques to the more problematic data sets typical of environmental monitoring and the paleorecord.

Keystone Species

Species whose removal results in a large effect on some functional property of an ecosystem—called *keystone species*—are good targets for indicators. Research is needed to develop a predictive theory of keystone species and to identify tolerant species. Traditionally, ecologists have looked for and identified keystone species by their effects on the species richness and composition of the community in which they live. Keystone species also may have major effects on primary and secondary productivity and nutrient cycling. Unfortunately, although a number of keystone species have been identified, no predictive theory of keystone species yet exists.

Tolerant Species and Assessing the Regional Importance of Local Sites

Evaluating places using only indicators that focus on each site separately would not lead to decisions that would sustain the greatest amount of species diversity. A hypothetical example suffices to explain why. Suppose a company intends to build a factory but it does not care on which of two equal-sized natural areas it builds. For purposes of preserving biological diversity, it might appear obvious that the site with the fewest species is preferable for the factory.

The issue is more complicated than that because one needs to know if the diversity on the richer site is sustainable (as has been taken into account for measuring species densities), and also the extent to which the site is redundant in the system of reserves that maintains S for the region as a whole. If the richer site has virtually the same list of species as another site and the species appear to be sustainable at the other site, the value of the rich site is lower. Conversely, if the alternate site for the factory, although relatively poor in diversity, harbors species found nowhere else in the region, or nowhere else in the world, it has higher biodiversity value than the richer site. The lessons of this simple example can be used to suggest ways to evaluate the contribution of a site, R_i, to regional diversity.

To achieve this, for example, separate S_i into two components. The species not found in sustainable condition elsewhere are R_i. The definition of "elsewhere" may vary depending on the objective of the group doing the evaluation. In other words, elsewhere means "anywhere else in the area whose diversity one is trying to protect." In this context, a species whose range is so fragmented that its survival depends on a number of sites should be counted as a full species in the R_i of each site that forms a necessary part of its support range.

To use R_i one must be able to measure sustainability. Sustainability is increased by avoiding overload in D_i, but understanding sustainability fully also requires knowledge of the population dynamics, metapopulation dynamics, and species interactions of the species contributing to R_i. Because obtaining all this knowledge is difficult and laborious, it cannot be done for all the species in a place. A reasonably accurate estimate of R_i can probably be obtained by focusing investigations on a few, charismatic taxa, such as birds, fishes, mammals, butterflies, wildflowers, and trees, instead of a random scattering of species in many taxa. However, considerable research will be needed to generate the data necessary for computing measures of sustainability.

As human uses take more of a region's area, many sites will shrink or disappear. As a result, the list of species restricted to a critical few sites and the average value of R_i will both climb. A place that lacks importance today may become important in the future. Thus, R_i must be reassessed periodically, probably once each decade. R_i is not an indicator but computing its values is an important part of an overall assessment of the performance of a system of reserves, so that biodiversity protection is achieved most efficiently.

5

Local and Regional Indicators

INTRODUCTION

Indicators are needed to inform us about ecological status and trends at all spatial and temporal scales, and at a variety of levels of specificity, ranging from the status of local populations to the functioning of large ecosystems. Because space and time are both continuous variables, scales of applicability of indicators blend into one another. Indeed, many indicators are useful at several scales. For example, the indicators we recommend in this chapter for forest condition can be aggregated usefully at regional, national, and continental scales.

In addition, most policy and management decisions are made at scales defined by laws and regulations established by political entities, such as local municipalities, counties, states, and the federal government. Although the committee focused its attention on the national-level ecological indicators recommended in Chapter 4, the methods used to select and formulate those indicators are equally applicable to indicators designed for use at smaller spatial scales. Further, many national-level indicators can be reported at various levels of disaggregation to serve as regional ecological indicators. In this chapter, we examine a number of local and regional indicators that we judge to be especially important, and show how they can be computed and interpreted.

PRODUCTIVITY INDICATORS

In addition to a national-level indicator of ecosystem productivity, it is also useful to have indicators specifically designed to capture the performance of particular ecosystem types. In this discussion, we give examples of indicators for forested ecosystems. Similar indicators can and should be developed for other vegetation types, such as grasslands, savannas, deserts, and wetlands.

FORESTS AS AN EXAMPLE

For regional forest indicators, we recommend indicators of productivity and species diversity, structural diversity, and sustainability. These attributes support the continued provision of the following goods and services from forests: wood and wood products, opportunities for recreation, tourism, and aesthetic enjoyment, maintenance of wildlife resources, control of erosion and nutrient losses to surface waters, and mitigation of greenhouse-gas emissions.

The most valuable indicators for forests are those that can provide early warning of adverse trends in productivity, species diversity, and structural diversity. Productivity integrates the flow and storage of carbon with flows of nutrients, water, and light. It provides the sustained yield of wood products. In addition, with the increased concentrations of carbon dioxide in the atmosphere, management of forests for carbon storage has assumed great importance (Cooper 1983, Harmon et al. 1990). Species diversity is also an important indicator of the condition of forests, if for no other reason than that most species on the Endangered Species List inhabit forests (Doyle 1998). Structural diversity of forests includes such features as crown condition, foliage-height profiles, and amounts and status of coarse woody debris; these attributes are all important for animal habitat (MacArthur et al. 1962; Franklin et al. 1981, 1989; Franklin and Forman 1987; Spies et al. 1988). The three features of forests that indicators address also provide opportunities for recreation and tourism and contribute towards maintaining the aesthetic quality of the nation's forests.

The development of a program for monitoring the status and trends of the nation's forested ecosystems is a continuing research effort that has sound practical underpinnings. Continuous inventory programs of the U.S. Forest Service form the basis of monitoring various aspects of forest structure and productivity. These inventory programs are positioned to take advantage of the existing theoretical base provided by individual-tree simulation models (Shugart 1984).

We first review the current forest inventory programs that provide the empirical basis for forest-status monitoring. We then discuss those current forest ecosystem models that may provide a theoretical basis for evaluating forest functioning, as well as for forecasting future status and trends in forest productivity and diversity. Based on this review, we then recommend specific indicators of the status of forests, focusing on how these indicators relate to current inventory programs and ecosystem models.

Current Forest Inventory Programs

The U.S. Forest Service has periodically inventoried the status of the nation's forests through its Forest Health Monitoring Program (FHMP) and the older Forest Inventory and Analysis (FIA) Program. A nationwide network of plots for the FHMP has been partially implemented (Anonymous 1996). In this program, forests are inventoried in plots spaced every 27 km on a nationwide grid network. The plot network currently exists in 15 states, mainly in the Northeast and Southeast, and in scattered areas in the rest of the country. At each grid point, various subplots are sampled for a variety of attributes. The size of each subplot is determined by the ecological scale of each attribute: the subplots range in area from 2 m^2 to 1 ha. These plots are sampled every four years for traditional timber-yield data on tree-diameter distributions, tree species, and site index. Canopy condition, leaf-area index, lichen communities, scenic beauty, lichen chemistry, foliar chemistry, dendrochemistry, dendrochronology, branch evaluation, browse supply, and root condition are also assessed (Anonymous 1996). The FHMP currently covers 70 percent of all forested lands in the coterminous United States. When fully implemented in 2002, the FHMP system will provide detailed data with reasonable spatial and temporal coverage to detect regional problems in the nation's forests.

The FHMP system provides detailed data on relatively few plots. In contrast, the FIA system provides extensive coverage on fewer attributes, mainly those related to timber volume and forest productivity. In the FIA program, permanent plots located on lands of all ownership types are inventoried every decade, mainly to evaluate standing crop of timber, but also in many cases for assessing understory vegetation, tree seedling regeneration, disease indicators, and browse availability. For example, in Minnesota alone, more than 10,000 plots have been inventoried in this manner since before 1960 in a cooperative program between the U.S. Forest Service and the Minnesota Department of Natural Resources. These data form the basis for policy decisions in Minnesota regarding timber supply from public lands and were the basis for the recently completed Generic Environmental Impact Statement for Expansion of the Pulp and Paper Industry in Minnesota (Jaakko-Pöyry Consulting 1992).

The FIA plots provide information on diameter, height, and species of all trees and numbers of seedlings by species. From these data, biomass, tree species diversity, and mortality can be calculated. On some plots, browse supply and condition are also measured to evaluate ungulate habitat. These FIA data can corroborate trends measured in more detail in the FHMP plots, and they provide more detailed coverage of the productivity and diversity of forested lands.

To be of even more use, the FIA system requires more complete data archiving and quality control. In particular, the locations of the FIA plots need to be determined accurately using global-positioning systems, rather than the current method of permanent stakes and survey markers, which are sometimes lost. Finally, additional FIA plots need to be established to reflect more accurately the distribution of land in various ownerships. For example, although the FIA plots in Minnesota are distributed across all ownership types, there are few FIA plots on national forest land in other states, particularly in the Pacific Northwest.

Current Simulation Models of Forest Ecosystems

The raw field data collected by the FHMP and FIA programs can be imported into individual tree-based ecosystem models (Shugart 1984) to project potential trends, given hypotheses about how various stressors (e.g., climate change, acid rain, and harvesting) affect tree physiology and stand population dynamics. The development of these ecosystem models in recent decades provides a theoretical basis for analysis and projection of the data (see examples reviewed in Ågren et al. 1991, Mladenoff and Pastor 1993, Pastor and Mladenoff 1993). These models project the diameter and height growth of individual trees on a plot approximately of 0.1 ha (approximately the same scale as the FIA and FHMP plots), subject to constraints of growth limitations, including light limitations through shading, temperature, water, and nutrients (Shugart 1984, Pastor and Post 1986, Pacala et al. 1996). The models have been tested against independent data on successional trends, productivity, species diversity, and nitrogen cycling throughout eastern North America (Shugart 1984, Pastor and Post 1986, Pastor et al. 1987, Pacala et al. 1996). Versions of these models also exist for the Pacific Northwest (Keenan et al. 1995). An example of using one of these models to determine plot sampling regimes for monitoring status and trends is detailed in Appendix A.

The combination of established, long-term monitoring programs and a suite of extensively tested simulation models operating at the same scale as the data provides a sound basis for an integrative program to assess the status and trends of the condition of the nation's forests. We now turn to specific recommendations for indicators that will enhance the usefulness of these models and inventory programs.

Recommended Indicators for the Status of the Nation's Forests

We recommend that the following forest indicators be given high priority: (1) productivity and tree species diversity, (2) soils, (3) light penetration, (4) foliage-height profiles, (5) crown condition, and (6) physical damage to trees. We recommend indicators that can be assessed with a small amount of time spent collecting data on site, that would be amenable to calculation of other synthetic indices (such as various diversity indices) later in the laboratory or office, and that could be easily incorporated into existing inventory programs.

1. *Productivity and Tree Species Diversity.* Productivity and tree species diversity form the basis of the forest food web; this web is sustained by the ability of soils to provide water and nutrients and by the ability of the crown to capture light. The FHMP and FIA programs already collect the data required to assess the status and trends of productivity and tree species diversity. These data include tree diameters at breast height, tree heights, density by species, height classes at which species occur, and canopy cover for each species within each height class. From these data, carbon storage and net primary productivity of trees can be calculated, as well as various diversity indices.

2. *Soils.* The soil profile should be characterized from a one-time sampling to characterize structure, texture, and rooting depth (Soil Survey Staff 1993). These physical features determine the ability of the soil to hold water at depths at which it is available to trees. Because these properties are relatively permanent, there is no need to reinventory them, except perhaps after several decades. In contrast to water-holding capacity, soil-nutrient availability can change fairly rapidly. Therefore, decadal sampling of the soil is required to determine changes in soil organic matter, total nitrogen, exchangeable cations, and forest floor chemistry (C/N ratio, lignin content, and P, K, Ca, and Mg contents). In addition, nitrogen availability should be assessed by means of resin bags (Binkley 1984). Doing so would require that the plots be revisited the following year to retrieve the bags, but this technique is able to assess rapid changes in availability of limiting nutrients (Pastor et al. 1998). Resin bag measurements therefore can serve as an early warning system of trends in soil productivity. The indicator of soil status (soil organic matter, recommended in Chapter 4) could enhance the soils component of the FHMP and FIA.

3. *Light Penetration.* Any disturbance to the canopy (or recovery of canopies from prior disturbance) necessarily changes light penetration. Light penetration can be measured easily by means of vertical photographs taken with a fish-eye lens, images that can then be analyzed later in a laboratory after they have been scanned into a computer. Furthermore, because changes in light penetration are the major driving force behind succession (Pacala et al. 1996), these data would be useful for projecting status and trends by means of the individual-tree models.

4. *Foliage-Height Profiles.* Foliage-height profiles are relatively easy to measure and provide a relative index of bird species diversity and possibly insect diversity. They are therefore an important measure of structural diversity. The vertical structure of the canopy, specifically foliage-height diversity, is strongly correlated with bird species diversity (MacArthur 1959, MacArthur et al. 1962, 1966). Foliage-height diversity varies temporally with canopy development during succession (Aber 1979a) and spatially along soil moisture gradients (Aber et al. 1982). This diversity can be measured rapidly by means of a camera that is mounted vertically and used as a range-finder (Aber 1979b). The vertical distribution of leaf area and therefore potential bird species diversity can then be calculated from these data. Such data are also useful for assessing changes in growth as well as its efficiency and allocation in forest stands (Ford 1982, Beadle et al. 1982, Waring 1982).

New technological developments may make it easier to collect such data. In particular, NASA has developed an instrument, the Vegetation Canopy Lidar (VCL), that can measure foliage-height profiles from above. The functioning prototype has operated successfully from the Space Shuttle and is now mounted on an aircraft. An updated version is being constructed for launch on a small satellite with a projected launch date of late 1999 or early 2000. This instrument will have a horizontal resolution of 20 m and is intended to map canopy heights over the entire surface of the planet once during its two-year lifetime. This goal may not be achieved, but to compare aerial or satellite imagery with stand data, extremely accurate spatial positions must be known for the stands, data that can be obtained only with global positioning systems.

5. *Crown Condition.* Crown condition, which reflects the state of the canopy that accounts for productivity, is correlated with tree energy status. Trees with a history of poor crowns usually grow more slowly and have diminished energy and carbon reserves. The latter characteristics translate into less carbon being available for defense against insects and pathogens and repair of damage caused by biotic or abiotic agents; energy reserves are critical in surviving periods of stress. Trends in damage and crown condition are usually accurate indicators of trends in productivity and mortality. If severe enough, damage and crown condition may

be useful predictors of mortality (e.g., Silver et al. 1991), although more general quantitative relationships have not been developed for those parameters.

Useful measurements of crown condition are crown ratios (percentage of tree height that supports live foliage), crown diameters, density, transparency, and dieback (progressive mortality of branches proceeding from branch tips inward). These data may be combined to give composite measures such as crown volume and crown surface area. Selected measurements, such as crown transparency and dieback, may also prove to be useful indicators. Continued evaluation may show how crown volume or crown surface area values can be useful indicators of habitat quality, especially for birds and insects.

6. *Physical Damage to Trees.* Trees are damaged by insects, pathogens, poor management practices, weather-related stresses, and air pollution, acting alone or in combination. Physical damage to trees after storms, lightning strikes, fire, and logging provides entry points for insect pests and diseases. The extent of damage and trends in damage, recorded by species and age class, can be diagnostic of cause in certain cases, and can provide a relative measure of likelihood that forest diversity, productivity, sustainability, and aesthetic value will be compromised. Combining quantitative measures of damage type and severity with mortality data could eventually provide a quantitative basis for predicting trends in valued forest attributes from damage indicators.

Damage categories used by foresters include wounds, evidence of pathogen attack, brooms, broken branches, broken roots, and damaged or discolored foliage, buds, and shoots. Weighting the components based on how likely the damage will effect growth or survival provides an index of the significance of the damage.

Implementation

The methods of collecting data on these indicators can be learned easily by field foresters in one- or two-day workshops. Most of the data processing would take place later in the laboratory through soil sample analysis or computer analysis. A mere 1 to 2 hours per plot are required to obtain these data (except for the initial, one-time soil profile characterization, which would require an additional 2 to 3 hours), beyond the time currently spent on collecting the more traditional timber evaluation data. Given the small increase in time spent in the field and the large advantage that would accrue from obtaining these data, such a program should receive high priority for development and implementation.

INDICATORS OF SPECIES DIVERSITY

In addition to the national indicators of the status of species diversity recommended in Chapter 4, the nation needs indicators to evaluate the diversity status of a local area, such as a national park or an area exploited for human use. For evaluating the diversity status of such areas, we recommend three indicators: *Independence of the Area*, *Species Density*, and *Deficiency of Natural Diversity*.

Although we tried to reduce the number of these indicators of diversity, and have grounded them in a single well-researched power law, all three are needed because they each inform us about different aspects of diversity. As Angermeier and Karr (1994) noted (about different levels of taxonomic diversity), "no accepted calculus permits integration of counts of elements across levels. . . . Arguably, no such calculus should be sought." We believe this point applies to diversity measures as well. It follows that the separate aspects of diversity need to be monitored and reported separately.

Local assessment of species diversity presents a new problem, because simple counts of species diversity have at least five weaknesses that make them unreliable.

- Diversity counts are biased by sample size (Fisher et al. 1943). The larger the sample, the more species in the count. Simple counts are rarely complete, and even when they are, one cannot be sure that they are. Moreover, the species most likely to be missed are the rarest—exactly those that most need protection.
- Diversity counts vary with the extent of the area over which they are measured. Larger areas have more species (Arrhenius 1921), not because they are environmentally better, but simply because they contain more habitats (Williamson 1943).
- Diversity counts are biased by the length of the period over which they are measured. More time leads to more species in the raw count (Preston 1960). Again, the greater number of species results not from improved environmental quality, but simply because longer durations yield greater heterogeneity. A longer period of observation is equivalent to more habitats in space (Rosenzweig 1998), because different species require various seasons and various kinds of years to succeed (Chesson 1994).
- Diversity is a dynamic property of ecosystems (Rosenzweig 1995). Simple counts do not tell us whether diversity is sustainable.
- The diversity of any area within a continent depends partly on the continental matrix in which that area is embedded (Rosenzweig 1995). Simple counts ignore that matrix. A species found in a place may persist there only because favorable conditions are accessible elsewhere.

Therefore, simple species counts need to be processed and analyzed before being incorporated into indicators. Fortunately, recent advances in the study of community diversity provide us with a number of sophisticated practical methods to minimize the sample-size bias (Burnham and Overton 1979, Chao 1987, Chazdon et al. 1998, Colwell and Coddington 1994). Our recommended local and regional indicators assume the use of these methods, and they also correct for the other deficiencies of simple counts.

In Chapter 4 we showed that samples of area contain a number of species, S, that fits a power law, $S = cA^z$, where A is area, and c and z are coefficients of the equation (Arrhenius 1921, Preston 1962). A thousand years ago, most of the sample areas monitored today were subsamples of a contiguous whole. They exhibited characteristic z values between 0.10 and 0.20 (Rosenzweig 1995). Now, however, they are likely to be isolated remnants of the whole, which is a very consequential difference for maintaining diversity.

Two types of species contribute to local diversity S (Shmida and Ellner 1984, Pulliam 1988, Pulliam and Danielson 1991). The first are species whose births exceed their deaths in the area. These are the *source species* of the sample. Other species, known as *sink species*, maintain themselves in a sample even though their average birth rate is less than their average death rate, because they frequently immigrate into the area.

Isolating an area, as usually happens when a reserve is set aside, cuts it off from the immigration that maintains its sink species (Rosenzweig 1995). The sink species then eventually vanish from the isolate. This reduces the c value and increases the z value characterizing the area's species diversity. The c value is idiosyncratic to particular taxa and regions, but z for isolates tends to be approximately 0.2 to 0.4, much higher than the z of subsamples (Rosenzweig 1995).

Knowing the relationship of S to area, the quasi-sustainable diversity of an area can be estimated by $cA^{0.3}$ (where c is the intercept coefficient of the regional, logarithmic species-area pattern for the taxon being assessed, and A is the size of the area being assessed.) Quasi-sustainable diversity is diversity that should persist for many human generations (although ultimately z rises and S declines, to a degree that is also predictable using the species-area relationship).

An Indicator of Independence

The quasi-sustainable S suggests an indicator of independence based on the z values most likely to characterize natural ecosystems. To compute this indicator, first assess the diversity of an area, S_i, and of its whole province, S_w. Let the area of interest be A_i and that of the province A_w. I_i, the measure of independence, is defined as:

$$I_i = [\log S_w - \log S_i] / 0.2[\log A_w - \log A_i]$$

The value 0.2 in the denominator is the threshold z value: $z > 0.2$ means no sink species. The rest of I_i is the z value of the area. Thus, if $I_i > 1$, the area probably contains few if any sink species and its diversity is independent of other parts of the province. If, on the other hand, $I_i < 1$, then some species living in the area rely on other areas for population support. These species need to be identified with more traditional demographic techniques. Their source areas need to be located and preserved as well.

An Indicator of Species Density

Managers typically wish to optimize the value of their reserves. It might appear that the more species housed in a reserve, the better its condition, but this is not necessarily true. As Chapter 4 showed, the form of the species-area curve means that an adjusted species density reveals more than raw species densities, S_i/A_i. Recall from Chapter 4 that the adjusted species density of an area is D_i, where $D_i = S_i/A_i^z$. The greater D_i, the more species the preserve maintains relative to the norm.

In calculating D_i, use the prevailing or average z value for the biological region. Because experience indicates that z is close to 0.3 in isolates, a value of 0.3 can be used if data are unavailable to estimate z. As in national assessments, high values of D_i must be interpreted carefully because they may reflect unsustainable overloading of the area. In particular, if high D_i is accompanied by $I_i < 1$, the high species density is unlikely to persist.

To see why this is true, consider an area that is not a reserve, but is used for various residential and commercial purposes. Despite this situation, suppose the area supports many wild species as well. If changing patterns of land use within the area squeeze those species into a more restricted, smaller proportion of the whole, D_i will rise and I_i will decline. If it is known that changing uses will remove a certain amount of the area from access by wildlife, the initial value of the higher D_i and lower I_i can be calculated in advance. But the increase in D_i does not signal environmental improvement because it is likely to decline to its former value. Once the effective area diminishes, the only way for the system to return to a sustainable diversity is through reduction in the actual number of species it contains. The increased D_i merely signals an impending loss of S_i.

We do not recommend that local diversity managers calculate the M value (see Chapter 4) of their areas instead of these areas' D value. At the local level, the indicator needs dictate whether the area is under- or overdiverse. M deliberately eliminates that distinction.

Indicators of Deficiency in Natural Diversity

When human uses dominate a landscape, natural assemblages of species disappear, but they are in part replaced by exotic species. In Chapter 4, we recommended a national indicator of native species diversity, to indicate the degree to which exotics have replaced native species. A local indicator that quantifies this tendency is also needed.

For example, consider the difference between the bird species of Tucson, Arizona, and those of the surrounding natural landscape (Emlen 1974, Table 5.1). The differences are typical of those seen in commensal assemblages of most or all other taxa, so we describe Emlen's conclusions and use them to design an index of deficiency in diversity, U_i.

Tucson sits in an Upper Sonoran Desert basin, surrounded by tracts of natural vegetation and their resident bird species. The city itself is mostly a vast suburb with expanses of vegetation supported by urban irrigation. As a result of extensive watering, the total abundance and biomass of all bird species has risen more than 26-fold, but most of the individuals belong to a few commensal and exotic species, mostly house sparrows, house finches, doves, starlings, and mockingbirds. Since Emlen's study, more curve-billed thrashers, cactus wrens, Gila woodpeckers, Gambel's quail, and pyrrhuloxias have moved into the city. Phainopeplas are more abundant in both the city and the desert. Anna's hummingbirds have virtually displaced the native black-chinned hummingbirds, and great-tailed grackles, a new commensal, have become quite common. Some raptors, such as Harris hawks, Cooper's hawks, and great horned owls, are now regularly seen in the city. But the overall difference Emlen observed has not changed in the two decades since he wrote: birds are far more abundant in Tucson than in the surrounding desert, but the city has fewer native species. Common, widespread opportunists and exotics account for most of the urban bird biomass.

There are several reasonable and complementary explanations for why anthropogenic habitats often bring about the loss of many native species and the burgeoning of commensals. First, anthropogenic habitats have no evolutionary pedigree; species have not had a chance to adapt to them. Moreover, people continue to change habitats at rates that are likely to prevent species from adapting to them. Nevertheless, a few species have traits that enable them to thrive in highly modified environments.

Many sets of species that use similar resources have members that depend for their continued mutual existence on their tolerance of suboptimal conditions. Tolerant species cannot dominate the "best" habitat patches, and intolerants depend on the best habitats for their survival. When people change a habitat to produce novel conditions, the most

TABLE 5.1 Emlen's study of the effect of urbanization on the avian species assemblage. Abundance went up 26-fold but diversity declined from 21 to 15 species. Moreover, many of the local specialties were replaced by exotics (house sparrow, starling) and widely distributed commensals like Inca doves, mockingbirds, cardinals and house finches.

	Individuals/100 acres	
Species	Desert	City
Gambel's quail	0.3	—
White-winged dove	0.5	140
Mourning dove	1.9	30
Inca dove	—	230
Roadrunner	0.5	—
Black-chinned hummingbird	—	6
Gilded flicker	1.9	—
Gila woodpecker	0.3	14
Ash-throated flycatcher	0.8	2
Verdin	2.5	14
Cactus wren	6.8	2
Curve-billed thrasher	6.9	5
Bendire's thrasher	0.2	—
Mockingbird	0.3	45
Black-tailed gnatcatcher	1.6	—
Starling	—	35
Loggerhead shrike	0.1	—
Brown-headed cowbird	0.4	1
Hooded oriole	0.6	—
House sparrow	—	520
Cardinal	—	17
Pyrrhuloxia	0.6	—
House finch	0.3	170
Brown towhee	1.2	—
Black-throated sparrow	16.5	—
Rufus-winged sparrow	2.5	—

tolerant species are likely to succeed exuberantly, whereas the intolerant ones become confined to nature reserves.

One reason why so many Old World species have moved in and exploded in suburban environments to the detriment of natives may be that they have had more time to adjust to humans. In addition, some species transplanted to new continents simultaneously escape from their predators. For example, an Australian native acacia that is not particularly abundant in the western Australian kwongan where it evolved, became a scourge in the similarly poor soils and Mediterranean climate of the similarly diverse fynbos in the Southwestern Cape Province of South Africa. Gypsy moth outbreaks are common in eastern North America, but rare in these insects' native Europe. (Diamond [1997] used similar ideas to understand the broad aspects of the distribution of human civilizations and the origin of technological advance.)

Thus, three factors contribute to the extraordinary abundance of a few species in anthropogenic environments:

- Exotics may have had more time to adjust to humans.
- Exotics may have escaped many of their natural predators.
- Only a subset of native species (the tolerants) are preadapted to "degraded" environments.

To evaluate the deficiency of diversity in an area of Tucson, one cannot use the raw value of D_i, species density, because it gives the city credit for exotic species that merely follow human settlement, and for tolerant natives that would thrive anywhere. An indicator that depreciates the value of an area according to the proportion of its species that thrive in anthropogenic habitats is needed. There is no shortage of such habitats, and there is not likely to be in the foreseeable future.

One way to achieve this would be to recalculate D_i, after excluding the contributions of the tolerants and the exotics. However, because not enough is currently known to identify tolerant species, the best that can be done is to exclude the exotics, as was done for NAT IMP (see Chapter 4). Let

$$G_i = S_{i,n}/cA^z,$$

where $S_{i,n}$ is the number of native species at the site and cA^z gives the number of species expected in a site of area A. (The coefficients c and z are determined for the taxon in areas of the region free of exotics. Therefore, cA^z amounts to an alternative expression for S_n.) Because G_i measures native species density, it makes a better index of local diversity and gives a truer picture of the value of a place in supporting diversity.

The complement of G_i is an indicator of true deficiency in diversity, labeled U to signify unnaturalness:

$$U_i = [cA^z - S_{i,n}]/cA^z.$$

U_i measures the proportion of native species expected at a site (of area A) but not found there. Thus, U_i combines the change due to exotic species with that caused by the overall loss of species. Because values of this indicator of the deficiency of natural diversity change relatively slowly with time, decadal monitoring is probably sufficient.

As an example, assuming that Emlen's data accurately reflect what to expect of 100 acres of Sonoran desert, one can calculate U_i for 100 acres of Tucson. In 100 acres, cA^z is 21 species. But $S_{i,n}$ is 12. Thus, U_i is 9/21 or 0.43. Tucson's 100 acres are 43 percent deficient in natural species diversity. Because everything gets scaled nonlinearly by the power law, that deficiency is the same for any amount of area in the city.

As another example of an indicator of deficiency of natural diversity, consider the proportion of nonnative fish species in a region, usually a state, because most states have agencies that collect data on the distribution and abundances of fishes, especially game fishes. New techniques are unlikely to change the ease of obtaining the necessary input data. The proportion of nonnative species in the fauna of a state is an imprecise measure because not all exotics are of equal importance. Deciding whether to list an exotic can be tricky. Some anadromous species that invade fresh waters only briefly for spawning are counted as exotics. Other nonnative species, such as guppies in Oregon, survive only in very specific environments, in this case, hot springs. Conversely, introduced trout cannot survive the hot summers in most lowland waters of the eastern United States, but they thrive in cool tailwaters below dams. Some nonnative fishes are temporary survivors that live for only a few weeks or months. Crude state lists do not distinguish such species from widespread permanent residents, but the methods described above can adjust an indicator so that very localized or temporary species do not count as much as other exotics.

The proportion of nonnative fish species varies greatly among states, from 0.02 (one [anadromous] species out of 56) in Alaska (Morrow 1980) to .47 (36 out of 76) for Washington State (Wydoski and Whitney 1979). In relatively dry western states with low diversities of native species and high proportions of dammed rivers, the indicator currently has values greater than 0.2, and for many states the values are greater than 0.3. In the wetter eastern states, which also have higher native fish diversities, values are about 0.07 to 0.08. Values in all states are almost certain to increase over time because established exotics are almost impossible to exterminate, new introductions continue, and the quality of habitat for native fishes continues to deteriorate in most states. Because most states have agencies that collect data on the distributions and abundances of fishes, especially game fishes, information is typically available by state rather than by region or ecoregion. Often, however, the information is old or not systematic, and so it is not always reliable, especially with respect to diagnosis and distribution of nonnative species.

The Index of Biotic Integrity: An Indicator of Species Diversity of Aquatic Ecosystems

Additive multimetric indicators have been developed and used to compare the species diversity of aquatic systems with what should be in those systems in the absence of human-caused perturbations (called *appropriate* diversity by the NRC [1994]). The most widely used multimetric indicator is the Index of Biotic Integrity (IBI), which has been developed and tested in a variety of aquatic ecosystems (Karr et al. 1986, Karr and Chu 1999). The use of IBI requires general agreement about which organisms indicate by their abundance or absence, poor or good ecological and water characteristics. The IBI provides a method for quantifying those qualitative assessments. The IBI is primarily a community-level rather than an ecosystem indicator because it is based on taxonomic assemblages within specific phylogenetic groups and specific biogeographic regions. The original IBI was developed for freshwater fish communities in streams in the Midwest. Recently, similar indicators have been applied to freshwater benthic macroinvertebrate communities in several regions (and even to some terrestrial communities).

An IBI is calculated from a set of measures of distribution and relative abundance of selected taxa. Each measure is assigned a numerical value—an integer ranging from 0 to 6—based on the qualitative judgment of the index developers. The final IBI, which is the sum of the individual scores (usually 10 to 12), is unbounded but typically is between 0 and 60. Because individual scores are discontinuous, statistical analysis of the additive scores are generally inappropriate. More detail on the mechanisms for developing multimetric indices was provided by Barbour et al. (1995).

Typically, IBIs are developed for biogeographic systems such as ecoregions where similar communities of organisms are expected. For example, Ohio has developed an extensive set of IBIs that vary by ecoregion, drainage area, and stream habitat (Ohio EPA 1987). As mentioned above, the statistical properties of additive indices make it unrealistic to add or average scores across spatial scales to create a national indicator (Norris 1995, Gerritsen 1995). The only effective way to aggregate measures into a multimetric indicator is already incorporated into the regulatory policies of the U.S. EPA and state environmental agencies. The Clean Water Act requires that attainment of water quality be reported on the basis of the number of stream miles meeting the criteria. As IBI scores have and will continue to be incorporated into state and federal regulatory standards, attainment will be reported in relation to stream miles assessed.

References

Aber, J. D. 1979a. Foliage-height profiles and succession in northern hardwood forests. Ecology 60:18-23.

Aber, J. D. 1979b. A method for estimating foliage-height profiles in broad-leaved forests. J. Ecology 67:35-40.

Aber, J. D., J. Pastor, and J. M. Melillo. 1982. Changes in forest canopy structure along a site quality gradient in southern Wisconsin. American Midland Naturalist 108:256-265.

Acevedo, M. F., D. L. Urban, and M. Ablan. 1995. Transition and gap models of forest dynamics. Ecological Applications 5:1040-1055.

Adamus, P. R., and K. Brandt. 1990. Impacts of Quality of Inland Wetlands of the United States: A Survey of Indicators, Techniques, and Applications of Community Level Biomonitoring Data. EPA/600/3-90/073. Environmental Protection Agency, Office of Research and Development, Washington, D.C.

Ågren, G. I., R. E. McMurtrie, W. J. Parton, J. Pastor, and H. H. Shugart. 1991. State-of-the-art of models of production-decomposition linkages in conifer and grassland ecosystems. Ecological Applications 1:118-138.

Allen, T. F. H., and T. B. Starr. 1982. Hierarchy: Perspectives from Ecological Complexity. University of Chicago Press, Chicago.

Alley, R. B., D. A. Meese, C. A. Schuman, A. J. Gow, K. C. Taylor, P. M. Grootes, J. W. C. White, M. Ram, E. D. Waddington, P. A. Mayewski, and G. A. Zielinski. 1993. Abrupt accumulation increase at the Younger Dryas termination in the GISP2 ice core. Nature 362:527-529.

Alves, D. S., and D. L. Skole. 1996. Characterizing land cover dynamics. International Journal of Remote Sensing 17:835-839.

Angermeier, P. L., and J. R. Karr. 1994. Biological integrity versus biological diversity as policy directives. BioScience 44:690-697.

Anonymous. 1996. Forest Health Monitoring. Forest Health Monitoring National Office, USDA, Research Triangle Park, NC.

Arbaugh, M., and L. Bednar. 1996. Statistical Considerations for Plot Design, Sampling Procedures, Analysis, and Quality Assurance of Ozone Injury Studies. USDA Forest Service General Technical Report PSW-gtr-155-29-34.

Arrhenius, O. 1921. Species and area. J. Ecol. 9:95-99.

Ayres, R. U., W. H. Schlesinger, and R. H. Socolow. 1994. Human impacts on the carbon and nitrogen cycles. Pp. 121-155 in R. H. Socolow, C. Andrews, F. Berkhout, and V. Thomas, eds. Industrial Ecology and Global Change. University Press, Cambridge, U.K.

Baker, L. A., and P. L. Brezonik. 1988. Dynamic model of internal alkalinity generation: calibration and application to precipitation-dominated lakes. Water Resour. Res. 24:65-74.

Baker, W. L. 1989. A review of models of landscape change. Landscape Ecology 2:111-113.

Ball, J. T., I. E. Woodrow, and J. A. Berry. 1986. A model predicting stomatal conductance and its contribution to the control of photosynthesis under different environmental conditions. Progress in Photosynthesis Research. Volume 4. J. Biggins, ed. Martinus-Nijhoff, Dordrecht, Netherlands. Pp. 221-224.

Band, L. E., D. L. Peterson, S. W. Running, J. C. Coughlan, R. Lammers, J. Dungan, and R. R. Nemani. 1991. Ecosystem processes at the watershed level: Basis for distributed simulation. Ecological Modeling 56:171-196.

Band, L. E., P. Patterson, R. R. Nemani, and S. W. Running. 1993. Forest ecosystem processes at the watershed scale: 2. Incorporating hillslope hydrology. Agricultural and Forest Meteorology 63:93-126.

Barbour, M. T., J. B. Stribling, and J. R. Karr. 1995. Multimetric Approach for Establishing Biocriteria and Measuring Biological Condition. In W. S. Davis and T. P. Simon, ed. Biological Assessment and Criteria: Tools for Water Resource Planning and Decision Making. Boca Raton, FL: Lewis Publishers.

Barlett, M. S. 1975. The Statistical Analysis of Spatial Pattern. Chapman and Hall, London.

Bauer, A., and A. L. Black. 1994. Quantification of the effect of soil organic matter content on productivity. Soil Sci. Soc. Am. J. 58:185-193.

Beadle, C. L., H. Talbot, and P. G. Jarvis. 1982. Canopy structure and leaf area index in a mature Scots pine forest. Forestry 55:105-123.

Beasley, D. B., and L. F. Huggins. 1982. ANSWERS—Areal Nonpoint Watershed Environmental Response Simulation: Users' Manual. USEPA document 905/9-82-001, U.S. Environmental Protection Agency, Chicago.

Beck, M. B. 1987. Water quality modeling: A review of the analysis of uncertainty. Water Resources Research 23:1393-1442.

Benke, A. C. 1990. Perspective on America's vanishing streams. J. North American Benthological Soc. 9:77-88.

Beven, K. J., ed. 1997. Distributed Models in Hydrology: Applications of TOPMODEL. John Wiley and Sons, Chichester, U.K.

Beven, K. J., and M. J. Kirkby. 1979. A physically based variable contributing area model of basin hydrology. Hydrological Sciences Bulletin 24:43-69.

Binkley, D. 1984. Ion exchange resin bags: Factors affecting estimates of nitrogen availability. Soil Science Society of America Journal 48:1181-1184.

Black, C. A. 1993. Soil fertility evaluation and control. Lewis Publ., Boca Raton, FL.

Bovee, K. B. 1996. Managing instream flow for biodiversity: A conceptual model and hypotheses. Pp. 83-100 in Proceedings of the Northern River Basins Study, NRBS Project Report No. 66, Edmonton, Alberta.

Bradshaw, G. A., and B. A. McIntosh. 1993. Detecting climate-induced change using wavelet analysis. Environmental Pollution 83(1 & 2):135-142.

Bradshaw, G. A., and T. A. Spies. 1992. Characterizing canopy gap structure in forests using the wavelet transform. Journal of Ecology 80:205-215.

Brezonik, P. L. 1994. Chemical Kinetics and Process Dynamics in Aquatic Systems. Lewis-CRC Press, Boca Raton, FL.

Broecker, W. S. 1987. Unpleasant surprises in the greenhouse? Nature 328:123-126.

Brown, C. D., D. E. Canfield, Jr., R. W. Bachman, and M. V. Hoyer. 1998. Seasonal patterns of chlorophyll, nutrient concentrations, and secchi disk transparency in Florida lakes. Lake and Reservoir Management 14:60-76.

Brown, D., R. Warwick, L. Cavalier, and M. Roller. 1977a. The persistence and condition of Douglas County, Minnesota lakes. Rept. No. 5021, Minn. Land Manag. Inform. System, Center for Urban and Regional Affairs, Univ. of Minnesota, Minneapolis, MN.

Brown, D. A., R. Warwick, and R. Skaggs. 1977b. Lake condition in east central Minnesota. Rept. No. 5022, Minn. Land Manag. Inform. System, Center for Urban and Regional Affairs, Univ. of Minnesota, Minneapolis, MN.

Brush, G. S. 1986. Geology and paleoecology of Chesapeake Bay: A long-term monitoring tool for management. Journal of the Washington Academy of Sciences (special volume in the Historical Perspective of the Estuary in Management) 76(3):146-160.

Brush, G. S., and F. W. Davis. 1984. Stratigraphic evidence of human disturbance in an estuary. Quaternary Research 18:91-108.

Buckley, J. 1986. Environmental effects of DDT: Case study. Pp. 358-374 in Ecological Knowledge and Environmental Problem Solving:Concepts and Case Studies. National Academy Press, Washington, D.C.

Burkholder, J. M. 1998. Implications of harmful microalgae and heterotrophic dinoflagellates in management of sustainable marine fisheries. Ecological Applications 8(1) Supplement:S37-S62.

Burnham, K. P., and W. S. Overton. 1979. Robust estimation of population size when capture probabilities vary among animals. Ecology 60:927-936.

Cairns, J., Jr., and J. R. Pratt. 1993. A history of biological monitoring using invertebrates. Pp. 10-27 in D. Rosenberg and V. Resh, eds., Freshwater Biomonitoring and Benthic Macroinvertebrates. Chapman and Hall, New York.

Carlson, R. E. 1977. A trophic state index for lakes. Limnol. Oceanogr. 22:361-369.

Carlson, R. E., and J. Simpson. 1996. A Coordinator's Guide to Volunteer Lake Monitoring Methods. North American Lake Management Society, Madison, WI.

Caswell, H. 1989. Matrix Population Models. Sinauer Associates, Inc. Sunderland, MA.

Chacon-Torres, A., L. G. Ross, and M. C. M. Beveridge. 1992. The application of SPOT multispectral imagery for the assessment of water quality in Lake Patzcuaro, Mexico. Int. J. Remote Sens. 13:587.

Chao, A. 1987. Estimating the population size for capture-recapture data with unequal catchability. Biometrics 43:783-791.

Chapra, S. C., and S. T. Tarapchak. 1976. A chlorophyll a model and its relationship to phosphorus loading plots for lakes. Water Resour. Res. 12:1260-1264.

Charles, D. F., J. P. Smol, and D. R. Engstrom. 1994. Paleolimnological approaches to biological monitoring. Pp. 233-293 in Biological Monitoring of Aquatic Systems, S.L. Loeb and A. Spacie, eds. CRC Press, Boca Raton, FL.

Chazdon, R. L., R. K. Colwell, J. S. Denslow, and M. R. Guariguata. 1998. Statistical methods for estimating species richness of woody regeneration in primary and secondary rain forests of NE Costa Rica. In: F. Dallmeier, J. A. Comiskey, eds. Forest Biodiversity Research, Monitoring and Modeling:Conceptual Background and Old World Case Studies. Parthenon, Paris.

Chesson, P. L. 1994. Multispecies competition in variable environments. Theoretical Population Biology 45:227-276.

Clark, J. S. 1988a. Drought cycles, the "little ice age" and fuels: A 750-year record of fire in northwestern Minnesota. Nature 334:233-235.

Clark, J. S. 1988b. Stratigraphic charcoal analysis on petrographic thin sections: Application to fire history in northwestern Minnesota. Quaternary Research 30:81-91.

Clark, J. S., J. Merkt, and H. Müller. 1989. Post-glacial fire, vegetation, and human history on the northern alpine forelands, south-western Germany. Journal of Ecology 77:897-925.

Cliff, A. D., and J. K. Ord. 1981. Spatial Processes: Models and Applications. Pion, London.

Cochrane, W. G. 1983. Planning and Analysis of Observational Studies. John Wiley and Sons, New York.

Cody, M. L. 1975. Towards a theory of continental diversities:bird distribution over Mediterranean habitat gradients. Pp. 214-257 in M. L. Cody and J. M. Diamond, eds., Ecology and Evolution of Communities. Belknap Press of Harvard Univ. Press, Cambridge, MA.

Cohen, Y., and J. Pastor. 1991. The responses of a forest ecosystem model to serial correlations of global warming. Ecology 72:1161-1165.

Collatz, G. J., J. A. Berry, and C. Grivet. 1992. Coupled photosynthesis-stomatal conductance model for leaves of C_4 plants. Australian Journal of Plant Physiology 19:519-538.

Colwell, R. K., and J. A. Coddington. 1994. Estimating terrestrial biodiversity through extrapolation. Philosophical Trans. Royal Society of London B 345:101-118.

Conquest, L. L. 1993. Statistical approaches to environmental monitoring: Did we teach the wrong things. Environmental Monitoring and Assessment 23(2-3):107-124.

Cooper, C. F. 1983. Carbon storage in managed forests. Canadian Journal of Forest Research 13:155-166.

Cooper, S. R., and G. S. Brush. 1991. Long-term history of Chesapeake Bay anoxia. Science 254:992-996.

Cowardin, L. M., V. Carter, F. C. Golet, and E. T. LaRoe. 1979. Classification of wetland and deepwater habitats of the United States. U.S. Fish and Wildlife Service, U.S. Department of the Interior, Washington, D.C.

Cramer, W., and C. B. Field. 1999. Comparing global models of terrestrial net primary productivity (NPP): Introduction. Global Change Biology 5 (Supplement):iii-iv.

Cramer, W., D. W. Kicklighter, A. Bondeau III, G. Churkina, B. Nemry, A. Ruimy, A. Schloss, J. Kaduk, and participants of the Potsdam NPP Model Intercomparison. 1999. Comparing global models of terrestrial net primary productivity (NPP): Overview and key results. Global Change Biology 5 (Supplement):1-15.

Cressie, N. A. C. 1993. Statistics for Spatial Data. John Wiley & Sons, New York.

Cude, C. 1996. Oregon Water Quality Index: Revision and Application. Department of Environmental Quality, Portland, OR.

Dahl, T. E., and C. E. Johnson. 1991. Wetland Status and Trends in the Conterminous United States Mid-1970s to Mid-1980s. U.S. Department of Interior, Fish and Wildlife Service, Washington, D.C.

Dai, A., and I. Y. Fung. 1993. Can climate variability contribute to the "missing" CO_2 sink? Global Biogeochemical Cycles 7:599-609.

Daniel, T. C., A. N. Sharpley, and J. L. Lemunyon. 1998. Agricultural phosphorus and eutrophication:A symposium overview. J. Environ. Quality 27:251-257.

David, M. B., L. E. Gentry, D. A. Kovacic, and K. M. Smith. 1997. Nitrogen balance in and export from an agricultural watershed. J. Environ. Qual. 26:1038-1048.

Davis, F. W. 1985. Historical evidence of submerged macrophyte communities of upper Chesapeake Bay. Ecology 66:981-993.

Davis, M. B. 1985. History of the vegetation on the Mirror Lake watershed. Pp. 53-65 in An Ecosystem Approach to Aquatic Ecology: Mirror Lake and Its Environment, G. E. Likens, ed. Springer-Verlag, New York.

Defries, R. S., and J. R. G. Townshend. 1994. NDVI-derived land-cover classifications at a global scale. International Journal of Remote Sensing 20(17):3567-3586.

Diamond, J. 1997. Guns, Germs, and Steel: The Fates of Human Societies. W. W. Norton and Company, New York.

Dillon, P. J., and F. H. Rigler. 1974a. The phosphorus-chlorophyll relationship in lakes. Limnol. Oceanogr. 19:767-773.

Dillon, P. J., and F. H. Rigler. 1974b. A test of a simple nutrient budget model predicting the phosphorus content of lake waters. J. Fish. Res. Bd. Canada 31:1771-1778.

Dixit, A. S., S. S. Dixit, and J. P. Smol. 1992. Algal microfossils provide high temporal resolution of environmental trends. Water, Air and Soil Pollution 62:75-87.

Dobson, A. P., J. P. Rodriguez, W. M. Roberts, and D. S. Wilcove. 1997. Geographic distribution of endangered species in the United States. Science 275:550-553.

Doljido, J. R., et al. 1994. Water quality index applied to rivers in the Vistula River Basin in Poland. Environmental Monitoring and Assessment 33:33-42.

Doyle, R. 1998. Forest density in the U.S. Scientific American 278:28.

Drake, J. A., H. A. Mooney, F. di Castri, R. H. Groves, F. J. Kruger, M. Rejmanek, and M. Williamson, eds. 1989. Biological Invasions. John Wiley & Sons, New York.

Elton, C. 1942. Voles, Mice, and Lemmings:Problems in Population Dynamics. Oxford University Press, London.

Emanuel, W. R., D. C. West, and H. H. Shugart. 1978. Spectral analysis of forest model time series. Ecological Modelling 4:313-326.

Emlen J. T., Sr. 1974. An urban bird community in Tucson, Arizona: Derivation, structure, regulation. Condor 76:184-197.

Engstrom, D. R., E. B. Swain, and J. C. Kingston. 1985. A paleolimnological record of human disturbance from Harvey's Lake, Vermont: Geochemistry, pigments, and diatoms. Freshwater Biology 15:216-288.

Environment Canada. 1991. The State of Canada's Environment. Ministry of Supply and Service, Ottawa.

Erwin, K. L. 1991. An Evaluation of Wetland Mitigation in the South Florida Management District, Volume 1. South Florida Management District, West Palm Beach.

FAO. 1994. Agro-ecological land resources assessment for agricultural development planning: A case study of Kenya. Making land use choices for district planning. World Soil Resource Reports 71/9. FAO, Rome, Italy.

Farquhar, G. D., S. von Caemmerer, and J. A. Berry. 1980. A biogeochemical model of photosynthetic CO_2 assimilation in leaves of C_3 species. Planta 149:78-90.

Fisher, R. A., A. S. Corbet, and C. B. Williams. 1943. The relation between the number of species and the number of individuals in a random sample of an animal population. J. Animal Ecology 12:42-58.

Foley, J. A., C. Prentice, N. Ramankutty, S. Levis, D. Pollard, S. Sitch, A. Haxeltine. 1996. An integrated biosphere model of land surface processes, terrestrial carbon balance, and vegetation dynamics. Global Biogeochemical Cycles 10(4):603-628.

Folke, C, J. Larsson, and J Sweitzer 1996. Renewable resource appropriation. Pp. 201-221 in: R. Costanza and O. Segura, eds. Getting Down to Earth. Island Press, Washington, D.C.

Forbes, S. A. 1887. The lake as a microcosm. Bulletin of the Peroria (Ill.) Science Association. Reprinted 1925 in the Bulletin of the Illinois Natural History Survey 15:537-550.

Ford, E. D. 1982. High productivity in a pole-stage Sitka spruce stand and its relation to canopy structure. Forestry 55:1-17.

Forest Stewardship Council. 1996. Web site http://www.fsc-uk.demon.co.uk/
Foster, D. 1995. Lane-use history and four hundred years of vegetation change in New England. Pp. 253-319 in Global Land Use Change: A Perspective from the Columbian Encounter, B. L. Turner, A. G. Sal, F. G. Bernaldez, and F. diCastri, eds. Madrid, Spain: Consejo Superior de Investigaciones Cientificas.
Franklin, J. F., and R. T. T. Forman. 1987. Creating landscape patterns by forest cutting: Ecological consequences and principles. Landscape Ecology 1(1):5-18.
Franklin, J. F., K. Cromack, Jr., W. Denison, A. McKee, C. Maser, J. Sedell, F. Swanson, and G. Juday. 1981. Ecological characteristics of old-growth Douglas-fir forests. USDA-Forest Service General Technical Report PNW-118, Pacific Northwest Forest and Range Experiment Station, Portland, OR.
Franklin, J. F., D. A. Perry, T. D. Schowalter, M. E. Harmon, A. McKee, and T. A. Spies. 1989. Importance of ecological diversity in maintaining long-term site productivity. Pp. 82-97 in D. A. Perry et al., eds. Maintaining the Long-term Productivity of Pacific Northwest Forests. Timber Press, Portland, OR.
Frayer, W. E. 1991. Status and Trends of Wetlands and Deepwater Habitats in the Conterminous United States, 1970s to 1980s. Michigan Technological University, Houghton.
Frost, T. M., D. L. DeAngelis, S. M. Bartell, D. J. Hall, and S. H. Hurlbert. 1988. Scale in the design and interpretation of aquatic community research. Pp. 229-258 in S. R. Carpenter, ed. Complex Interactions in Lake Communities. Springer-Verlag, New York.
Fung, J. C. J. Tucker, and K. C. Prentice. 1987. Application of very high resolution radiometer vegetation index to study atmosphere-biosphere exchange of CO_2. Journal of Geophysical Research 92:2999-3015.
Gallie, E. A., and P. A. Murtha. 1993. A modification of chromaticity analysis to separate the effects of water quality variables. Remote Sens. Environ. 44:47-65.
Gerritsen, J. 1995. Additive biological indices for resource management. JNABS 14(3):451-457.
Gorham, E. 1995a. The biogeochemistry of northern peatlands and its possible responses to global warming. Pp. 169-187 in Biotic Feedbacks in the Global Climate System. G. M. Woodwell and F. T. Mackenzie, eds. Oxford University Press, New York.
Gorham, E. 1995b. Wetlands: An essential component of curricula in limnology. Pp. 234-246 in Freshwater Ecosystems: Revitalizing Education in Limnology. National Academy Press, Washington, D.C.
Gorham, E. 1991. Northern peatlands role in the carbon cycle and probable responses to climatic warming. Ecological Applications 1:182-195.
Government of the Netherlands. 1991. National Environmental Program, 92-95. Second Chamber of Parliament, The Hague.
Graumlich, L. J. 1993. A 1000-year record of temperature and precipitation in the Sierra Nevada. Quaternary Research 39:249-255.
Greenwalt, L. A. 1992. Global indicators: What the people expect. Pp. 109-114 in D. H. McKenzie, D. E. Hyatt, and V. J. McDonald, eds. Ecological Indicators, Volume 1. Elsevier Applied Science, London and New York.
Grime, J. P. 1997. Biodiversity and ecosystem function: The debate deepens. Science 277:1260-1261.
Grootes, P. M., M. Stuiver, J. W. C. White, S. J. Johnson, and J. Jouzel. 1993. Comparison of the oxygen isotope records from the GISP2 and GRIP Greenland ice cores. Nature 366:552-554.
Gunderson, L. H., C. S. Holling, and S. S. Light. 1995. Barriers and Bridges to the Renewal of Ecosystems and Institutions. Columbia University Press, New York.

Hall, D. K., G. A. Riggs, and V. V. Salomonson. 1995. Development of methods for mapping global snow cover using moderate resolution imaging spectroradiometric data. Remote Sens. Environ. 54:127.

Hall, F. G., D. B. Botkin, D. E. Strebel, K. D. Woods, and S.J. Goetz. 1991. Large-scale patterns of forest succession as determined by remote sensing. Ecology 72:628-640.

Hammond, A., A. Adriaanse, E. Rodenburg, D. Bryant, and R. Woodward. 1995. Environmental Indicators: A Systematic Approach to Measuring and Reporting on Environmental Policy Performance in the Context of Sustainable Development. World Resources Institute, Washington, D.C.

Harmon, M. E., W. K. Ferrell, and J. F. Franklin, 1990. Effects on carbon storage of conversion of old-growth forests to young forests. Science 247:699-701.

Harris, B. L., T. L. Nipp, D. K. Waggoner, and A. Lueber. 1995. Agricultural water quality program policy considerations. J. Environ. Qual. 24:405-411.

Haxeltine, A., and I. Colin Prentice. 1996. BIOM3: An equilibrium terrestrial biosphere model based on ecophysiological constraints, reosurce availability, and competition among plant functional types. Global Biogeochemical Cycles 10(4):693-709.

Heinselman, M. L. 1996. The Boundary Waters Wilderness Ecosystem. University of Minnesota Press, Minneapolis.

Heiskary, S. A., and W. W. Walker, Jr. 1988. Developing phosphorus criteria for Minnesota lakes. Lake and Reservoir Management 4:1-10.

Hilsenhoff, W.L. 1987. An improved biotic index of organic stream pollution. Great Lakes Ent. 20:31-39.

Hilsenhoff, W.L. 1982. Using a biotic index to evaluate water quality in streams. Wisc. Dept. Nat. Res. Tech. Bull. 132:1-22.

Horn, H. S. 1975. Markovian processes of forest succession. Pp. 196-211 in M. L. Cody and J. M. Diamond, eds. Ecology and Evolution of Communities. Belknap Press, Cambridge, MA.

Howarth, R. W., G. Billen, D. Swaney, A. Townsend, N. Jaworski, K. Lajtha, J. A. Downing, R. Elmgren, N. Caraco, T. Jordan, F. Berendse, J. Freney, V. Kudeyarov, P. Murdoch, and Z. Zhao-Liang. 1996. Regional nitrogen budgets and riverine N & P fluxes for the drainages to the North Atlantic Ocean:Natural and human influences. Biogeochem. 35:75-139.

Hubbard, B. B. 1996. The World According to Wavelets. A. K. Peters. Wellesley, MA.

Hughes, M. K., and L. J. Graumlich. 1996. Multimillennial dendroclimate records from the western United States. Pp. 109-124 in Climatic Variations and Forcing Mechanisms of the Last 2000 years. In R. S. Bradley, P. D. Jones, and J. Jouzel, eds. Springer Verlag, New York.

Hurtt, G. C., and R. A. Armstrong. 1996. A pelagic ecosystem model calibrated with BATS data. Deep-Sea Research II 43:653-683

Hurtt, G. C., P. R. Moorcroft, S. W. Pacala, and S. Levin. 1998. Terrestrial models and global change. Challenges for the future. Global Change Biology 4:581-590.

HydroQual, Inc. 1998. Advanced eutrophication model of the Upper Mississippi River. Task 3. Prepared for Metropolitan Council Environmental Services, St. Paul, MN. HydroQual, Inc., Mahwah, NJ. (draft)

HydroQual, Inc. 1991. Water quality modeling analysis of hypoxia in Long Island Sound. Prepared for Management Committee Long Island Sound Estuary Study and New England Interstate Water Pollution Control Commission. HydroQual, Inc., Mahwah, NJ.

IGBP. 1995. The IGBP Terrestrial Transects. IGBP Report 34, IGBP Secretariat, Stockholm.

IGBP. 1992a. Global change and terrestrial ecosystems: The Operational Plan. IGBP Report 21, IGBP Secretariat, Stockholm.

IGBP. 1992b. Improved global data for land applications: A proposal for a new high resolution data set. J. R. G. Townshend, ed. IGBP Report 20, IGBP Secretariat, Stockholm.

IPCC. 1995. Second Assessment Report. Climate Change 1995: The Science of Climate Change. Cambridge University Press, U.K.

Jaakko-Pöyry Consulting. 1992. Generic Environmental Impact Statement for Expansion of the Pulp and Paper Industry in Minnesota. Report for Environmental Quality Board, St. Paul, Minn.

Janetos, A. et al. 1997. Workshop Report. CEOS Pilot Project:Global Observations of Forest Cover (GOFC). Ottawa, Ontario, Canada, July 7-10, 1997.

Jarvinen, O. 1985. Conservation indices in land use planning:Dim prospects for a panacea. Orins Fennica 62:101-106.

Jenkinson, D. S., and K. A. Smith, eds. 1988. Nitrogen Efficiency in Agricultural Soils. Elsevier Applied Sci. Publ., London.

Jenny, H. 1941. Factors of Soil Formation. McGraw-Hill Book Company, Inc., New York.

Jin, K. R., R. T. James, W. S. Lung, D. P. Loucks, R. A. Park, and T. S. Tisdale. 1998. Assessing Lake Okeechobee eutrophication with water-quality models. J. Water Resour. Plan. Manage. 124:22-30.

Johnson, D. W., and S. E. Lindberg, eds. 1992. Atmospheric Deposition and Forest Nutrient Cycling. Springer-Verlag, New York.

Johnston, C. A., and R. J. Naiman. 1990. The use of geographic information systems to analyze long-term landscape alteration by beaver. Landscape Ecology 4:5-19.

Justice, C. O., and J. R. G. Townshend. 1988. Selecting the spatial resolution of satellite sensors required for global monitoring of land transformations. International Journal of Remote Sensing 9(2):187-236.

Justice, C. O., J. P. Malingreau, and A. Setzer. 1993. Satellite remote sensing of fires:Potential and limitation. Pp. 77-88 in P. Crutzen and J. Goldammer, eds. Fire in the Environment: The Ecological, Atmospheric, and Climatic Importance of Vegetation Fires. John Wiley and Sons, Chichester, U.K.

Justice, C. O., J. D. Kendall, P. R. Dowty, and R. J. Scholes. In press. Satellite remote sensing of fires during the SAFARI Campaign using NOAA-AVHRR data. Journal of Geophysical Research.

Karr, R. J. 1996. Ecological integrity and ecological health are not the same. Pp. 97-109 in P. C. Schulze, ed. Engineering Within Ecological Constraints. National Academy Press, Washington, DC.

Karr, R. J., and E. W. Chu. 1999. Restoring Life in Running Waters: Better Biological Monitoring. Island Press, Washington, DC.

Karr, R. J., L A. Toth, and D. R. Dudley. 1985. Fish communities of midwestern rivers: A history of degradation. Bioscience 35:90-95.

Karr, R. J., K. D. Fausch, P. L. Angermeier, P. R. Yant, and I. J. Schlosser. 1986. Assessment of biological integrity in running waters: A method and its rationale. Illinois Natural History Survey Special Publication No. 5. Illinois Natural History Survey, Champaign, IL.

Kates, R. W., and W. C. Clark. 1996. Environmental surprise: Expecting the unexpected. Environment 38:6-34.

Keenan, R. J., J. P. Kimmins, and J. Pastor. 1995. Modeling carbon and nitrogen dynamics in western red cedar and western hemlock forests. Pp. 547-568 in Carbon Forms and Functions in Forest Soils. Proceedings from the 8th North American Forest Soils Conference, Gainesville, FLa. Soil Science Society of American, Madison, WI.

Keeney, D. R., and T. H. DeLuca. 1993. Des Moines River nitrate in relation to watershed agricultural practices:1945 versus 1980s. J. Environ. Qual. 22:267-272.

Keith, L. B. 1963. Wildlife's Ten Year Cycle. University of Wisconsin Press, Madison, WI.

Kelly, C. A., J. W. M. Rudd, R. H. Hesslein, D. W. Schindler, P. J. Dillon, C. T. Driscoll, S. A. Gherini, and R. E. Hecky. 1988. Prediction of biological acid neutralization in acid-sensitive lakes. Biogeochemistry 3:129-140.

Kenkel, N. C. 1993. Modeling Markovian dependence in populations of *Aralia nudicalis*. Ecology 74:1700-1706.

Kidwell, K. B. 1997. NOAA Global Vegetation Index User's Guide, July 1997 Revision. U.S. Department of Commerce, National Oceanic and Atmospheric Administration, National Environmental Satellite, Data, and Information Service, Washington D.C. http://www2.ncdc.noaa.gov/docs/.

Kirchoff, V. W. J., II. 1994. TAHBIS – A tropical atmosphere-hydrosphere-biosphere integrated study in the Amazon. Brazilian Journal of Geophysics (Revista Brasilera de Geofisica) 12(1):3-8.

Kloiber, S., T. Anderle, P. L. Brezonik, L. Olmanson, M. Bauer, and D. Brown. In press. Satellite imagery: An efficient approach to regional monitoring of lake trophic state conditions. Arch. Hydrobiol. Ergebn. Limnol.

Kolkwitz, R., and M. Marsson. 1909. Oekologie der tierischen Saproben. Int. Rev. Gesamten Hydrobiol. Hydrogr. 2:126-152.

Kolpin, D. 1997. Agricultural chemicals in groundwater of the midwestern U.S.:Relations to land use. J. Environ. Quality 26:1025-1037.

Kratz, T. K. 1995. Temporal and spatial variability as neglected ecosystem properties: Lessons learned from 12 North American Ecosystems. Conference on Evaluating and Monitoring the Health of Large-Scale Ecosystems. D. J. Rapport, C. Gaudet, and P. Calow, eds. NATO ASI Series Vol. 28:359-384. Springer-Verlag, Berlin, Heidelberg.

Kratz, T. K., B. J. Benson, E. Blood, G. L. Cunningham, and R. A. Dahlgren. 1991. The influence of landscape position on temporal variability in four North American ecosystems. American Naturalist 138:355-378.

Kusler, J. A., and M. E. Kentula. 1989. Wetland Creation and Restoration: The Status of the Science. Volume 1: Regional Reviews. Volume 2: Perspectives. Report 600/3-89/038. U.S. Environmental Protection Agency Environmental Research Laboratory, Corvallis, OR.

Landres, P.B. 1992. Ecological indicators: Panacea or liability? Pp. 1295-1318 in D. H. McKenzie, D. E. Hyatt, and V. J. McDonald, eds. Ecological Indicators, Volume 2. Elsevier Applied Science, London and New York.

Laporte, N., C. O. Justice, and J. Kendall. 1995. Mapping the humid forests of central Africa using NOAA-AVHRR data. International Journal of Remote Sensing 16(6):1127-1145.

Larson, W. E., and F. J. Pierce. 1991. Conservation and enhancement of soil quality. Pp. 175-203 in Evaluation for Sustainable Land Management in the Developing World, Vol. 2. IBSRAM Proc. 122 Technical Papers. Thailand: International Board for Soil Research and Management, Bangkok.

Lasaga, A. C. 1985. The kinetic treatment of geochemical cycles. Geochim. Cosmochim. Acta 44:815-828.

Lathrop, R. G. 1992. Landsat thematic mapper monitoring of turbid inland water quality. Photogram. Engrg. Remote Sens. 58:465.

Leith, H. 1972. Modeling the primary productivity of the world. Nature and Resources 8(2):5-10.

Leitner, W. A., and M. L. Rosenzweig. 1997. Nested species – area curves and stochastic sampling:A new theory. Oikos 79:503-512.

Letourneau, D. K. 1997. Plant-arthropod interaction in agroecosystems. Pp. 239-290 in Ecology in Agriculture, L.E. Jackson, ed. Academic Press, San Diego, CA.

Lichtenberg, E., and L. K. Shapiro. 1997. Agriculture and nitrate concentrations in Maryland community water system wells. J. Environ. Qual. 26:145-153.

Lillesand, T. M., W. L. Johnson, R. L. Deuell, O. M. Lindstrom, and D. E. Meisner. 1983. Use of Landsat data to predict the trophic state of Minnesota lakes. Photogram. Engrg. Remote Sens. 49:219-229.

Lindeman, R. L. 1942. The trophic-dynamic aspect of ecology. Ecology 23:399-418.

Lindsey, S. D., R. W. Gunderson, and J. P. Riley. 1992. Spatial distribution of point soil moisture estimates using Landsat TM data and Fuzzy-C classification. Water Resour. Bulletin 28:865.

Linthurst, R. A., D. H. Landers, J. M. Eilers, P. E. Kellar, D. F. Brakke, W. S. Overton, R. Crowe, E. P. Meier, P. Kanciruk, and D. S. Jeffries. 1986. Regional chemical characteristics of lakes in North America. II. Eastern United States. Water, Air, and Soil Pollution 31:577-591.

Lippe, E., J. Y. DeSmidt, and D. C Glenn-Lewin. 1985. Markov models and succession:A test from a heathland in the Netherlands. J. Ecology 73:775-791.

Lloyd, A. H., and L. J. Graumlich. 1997. Holocene dynamics of treeline forests in the Sierra Nevada. Ecology 78:1199-1210.

Lluch-Belda, D., R. J. M. Crawford, T. Kawasaki, A. D. MacCall, R. H. Parrish, R. A. Schwartzoise, and P. E. Smith. 1989. World-wide fluctuations of sardine and anchovy stocks: The regime problem. South African J. of Marine Science 8:195-205.

Los, S. O., G. J. Collatz, P. J. Sellers, C. M. Malmstrom, N. H. Pollack, R. S. DeFries, C. J. Tucker, L. Bounova, and D. A. Dazlich. In press. A global 9-year biophysical landsurface dataset from NOAA AVHRR data. Journal of Geophysical Research.

Lozano-Garcia, D. F., and R. M. Hoffer. 1985. Evaluation of a layered approach for classifying multi-temporal Landsat MSS. Pp. 189-199 in Pecora 10: Remote Sensing in Forest and Range Resource Management: Proceedings, August 20-21, 1985, Student Center, Colorado State University, Fort Collins, Colorado. William T. Pecora Memorial Symposium on Remote Sensing. American Society for Photogrammetry and Remote Sensing, Falls Church, VA.

Lutz, E., and S. El-Serafy. 1989. Environmental and Resource Accounting: An Overview. The World Bank, Washington, DC.

Lyche, A., T. Andersen, K. Christofferen, D. O. Hessen, P. H. Berger Hansen, and A. Klysner. 1996. Mesocosm tracer studies. 1. Zooplankton as sources and sinks in the pelagic phosphorus cycle of a mesotrophic lake. Limnol. Oceanogr. 41:460-474.

MacArthur, R. H. 1959. On the breeding distribution pattern of North American migrant birds. Auk. 75:318-325.

MacArthur, R. H., J. W. MacArthur, and J. Preer. 1962. On bird species diversity. II. Prediction of bird census from habitat measurements. American Naturalist 96:167-174.

MacArthur, R. H., H. Recher, and M. L. Cody. 1966. On the relation between habitat selection and bird species diversity. American Naturalist 100:319-332.

MacDonald, G. M., T. W. D. Edwards, K. A. Moser, R. Pienitz, and J. P. Smol. 1993. Rapid response of treeline vegetation and lakes to past climate warming. Nature 361:243-246.

Magdoff, F., L. Lanyon, and B. Liebhardt. 1997. Nutrient cycling, transformation and flows:Implications for a more sustainable agriculture. Advances Agron. 60:1-71.

Martin, M. E., and J. D. Aber. 1997. High spatial resolution remote sensing of forest canopy lignin, nitrogen, and ecosystem processes. Ecological Applications 7:431-443.

Matson, P. A., W. J. Parton, A. G. Power, and M. J. Swift. 1997. Agricultural intensification and ecosystem properties. Science 277:504-509.

McAndrews, J. H., and M. Boyko-Diakonow. 1989. Pollen analysis of varved sediment at Crawford Lake, Ontario: Evidence of Indian and European farming. Pp. 528-530 in Quaternary Geology of Canada and Greenland, R. J. Fulton, ed. Geological Survey of Canada, Ottawa, Ontario.

Metcalfe, J. L. 1989. Biological water quality assessment of running waters based on macroinvertebrate communities: History and present status in Europe. Environmental Pollution Series A 60:101-139.

Meyer, W. B., and B. L. Turner III, eds. 1994. Changes in Land Use and Land Cover: A Global Perspective. Cambridge University Press, Cambridge, U.K.

Mitsch, W. J., and J. G. Gosselink. 1993. Wetlands, 2nd ed. Van Nostrand Publishing Company, New York.

Mittenzwey. K.-H., S. Ullrich, A. A. Gitelson, and K. Y. Kondratiev. 1992. Determination of chlorophyll *a* of inland water on the basis of spectral reflectance. Limnol. Oceanogr. 37:147-149.

Mladenoff, D. J., and J. Pastor. 1993. Sustainable forest ecosystems in the northern hardwood and conifer region:Concepts and management. Pp. 145-180 in G. H. Aplet, J. T. Olson, N. Johnson, and V. A. Sample, eds. Defining Sustainable Forestry. Island Press and The Wilderness Society, Washington, DC.

Mooney, H. A., J. H. Cushman, E. Medina, O. E. Sala, and E.-D. Schulze. 1996. Functional Roles of Biodiversity: A Global Perspective. John Wiley & Sons, New York.

Morrow, J. E. 1980. The Freshwater Fishes of Alaska. Alaska Northwest Publishing Company, Seattle, WA.

Myneni, R. B., C. D. Keeling, C. J. Tucker, G. Asrar, and R. R. Nemani. 1997. Increased plant growth in the northern high latitudes from 1981-1991. Nature 386:698-702.

Naeem, S., L. J. Thompson, S. P. Lawler, J. H. Lawton, and R. M. Woodfin. 1994. Declining biodiversity can alter the performance of ecosystems. Nature 368:734-737.

National Research Council (NRC). 1986. Ecological Knowledge and Environmental Problem-Solving: Concepts and Case Studies. National Academy Press, Washington D.C.

National Research Council (NRC). 1992. Assessment of the U.S. Outer Continental Shelf Environmental Studies Program. II. Ecology. National Academy Press, Washington, D.C.

National Research Council (NRC). 1993. Soil and Water Quality:An Agenda for Agriculture. National Academy Press, Washington, D.C.

National Research Council (NRC). 1994. Review of EPA's Environmental Monitoring and Assessment Program: Surface Waters. National Academy Press, Washington, D.C.

National Research Council (NRC). 1995a. Review of EPA's Environmental Monitoring and Assessment Program: Overall Evaluation. National Academy Press, Washington, D.C.

National Research Council (NRC). 1995b. Wetlands: Characteristics and Boundaries. National Academy Press, Washington, D.C.

National Research Council (NRC). 1996a. The Bering Sea Ecosystem. National Academy Press, Washington, D.C.

National Research Council (NRC). 1996b. Freshwater Ecosystems: Revitalizing Educational Programs in Limnology. National Academy Press, Washington, D.C.

National Research Council (NRC). 1999a. Sustaining Marine Fisheries. National Academy Press, Washington, DC.

National Research Council (NRC). 1999b. Water for the Future: The West Bank and Gaza Strip, Israel, and Jordan. National Academy Press, Washington, DC.

Neilson, R. P. 1995. A model for predicting continental-scale vegetation distribution and water balance. Ecological Applications 5:362-385.

Nicholson, S., C. J. Tucker, and M. B. Ba. 1998. Desertification, drought, and surface vegetation: An example from the West African Sahel. Bulletin of the American Meteorological Society 79:815-831.

NMFS (National Marine Fisheries Service). 1996. Our Living Oceans:Report on the Status of U.S. Living Marine Resources 1995. NOAA Technical Memorandum NMFS – F/SPO-19. National Marine Fisheries Service, Silver Spring, MD.

Norris, R. H. 1995. Biological monitoring: The dilemma of data analysis. JNABS 14(3):440-450.

Nyquist, H. 1928. Certain topics in telegraph transmission theory. Transactions of the American Institute of Electrical Engineering 47:617-644.

Odum, E. P. 1984. Properties of Agroecosystems. Pp. 3-12 in Agricultural Ecosystems: Unifying Concepts. R. Lowrance, B. R. Stinner, and G. J. House, eds. Wiley Interscience, New York.

Ohio EPA. 1987. Biological Criteria for the Protection of Aquatic Life. Volume I, II, III. Ohio EPA Division of Water Quality Monitoring and Assessment, Surface Water Section, Columbus, OH.

Overpeck, J. T. 1996. Warm climate surprises. Science 271:1820-1821.

Pacala, S. W., C .D. Canham, J. Saponara, J. A. Silander, R. K. Kobe, and E. Ribbens. 1996. Forest models defined by field measurements: Estimation, error analysis and dynamics. Ecological Monographs 66:1-43.

Palacios-Orueta, A. and S. L. Ustin. 1996. Multivariate statistical classification of soil spectra. Remote Sens. Environ. 57:108.

Parker, R. A. 1968. Simulation of an aquatic ecosystem. Biometrics 24:803-821.

Parr, J. F., R. I. Papendick, S. B. Hornick, and R. E. Meyer. 1992. Soil quality:Attributes and relationship to alternative and sustainable agriculture. Amer. J. Alter. Agriculture 7:5-11.

Parton, W. J., W. M. Pulliam, and D. S. Ojima. 1994. Application of the CENTURY model across the LTR network: Parameterization and climate change simulations. Bulletin of the Ecological Society of America 75:186-194.

Parton, W. J., A. N. Scott, D. S. Ojima, R. McKane, and E. B. Rastetter. 1992. Carbon storage in terrestrial ecosystems: A comparison of the CENTURY and GEM ecosystem simulation models. Bulletin of the Ecological Society of America 73:340.

Parton, W. J., J. Stewart, and C. Cole. 1988. Dynamics of C, N, P and S in grassland soils: A model. Biogeochemistry 5:109-131.

Pastor, J., and Y. Cohen. 1997. Nitrogen cycling and the control of chaos in a boreal forest model. Pp. 304-319 in K. Judd, A. Mees, K. L. Teo, and T. L. Vincent, eds. Control and Chaos. Birkhaüser, Boston.

Pastor, J., and D. J. Mladenoff. 1993. Modelling the effects of timber mangement on population dynamics, diversity, and ecosystem processes. Pp. 16-29 in D. C. Le Master and R. A. Sedjo, eds. Modelling Sustainable Forest Ecosystems. American Forests, Washington, DC.

Pastor, J., and W. M. Post. 1993. Linear regressions do not predict the transient responses of eastern North American forests to CO_2 induced climate change. Climatic Change 23:111-119.

Pastor, J., and W. M. Post. 1988. Response of northern forests to CO_2-induced climatic change. Nature 334:55-58.

Pastor, J., and W. M. Post. 1986. Influence of climate, soil moisture, and succession on forest carbon and nitrogen cycles. Biogeochemistry 2:3-27.

Pastor, J., R. H. Gardner, V. H. Dale, and W. M. Post. 1987. Successional changes in soil nitrogen availability as a potential factor contributing to spruce dieback in boreal North America. Canadian Journal of Forest Research 17:1394-1400.

Pastor, J., J. Bonde, C. A. Johnston, and R. J. Naiman. 1992. Markovian analysis of the spatially dependent dynamics of beaver ponds. Lectures on Mathematics in the Life Sciences 23:5-27.

Pastor, J., B. Dewey, R. Moen, M. White, D. Mladenoff, and Y. Cohen. 1998. Spatial patterns in the moose-forest-soil ecosystem on Isle Royale, Michigan, USA. Ecological Applications 8:411-424.

Patrick, R., J. Cairns, Jr., and S. Roback. 1967. A study of the numbers and kinds of species found in rivers of the eastern United States. Proc. Acad. Nat. Sci., Phila., PA.

Pauly, D. 1995. Anecdotes and the shifting baseline syndrome of fisheries. TREE 10(10):430.

Paustian, K., H. P. Collins, and E. A. Paul. 1997. Management controls on soil carbon. Pp. 15-49 in E. A. Paul, E. T. Elliot, K. Paustian, and C. V. Cole, eds. Soil Organic Matter in Temperate Agroecosystems. CRC Press, Boca Raton, FL.

Peterjohn, B. J., J. R. Sauer, and S. Orsillo. 1995. Breeding bird survey: Population trends 1966-92. Pp. 17-21 in E. T. LaRoe, G. S. Farris, C. E. Puckett, P. D. Doran, and M. J. Mac, eds. Our Living Resources: A Report to the Nation on the Distribution, Abundance, and Health of U.S. Plants, Animals, and Ecosystems. U.S. Department of the Interior, National Biological Service, Washington, DC.

Peterson, D. L., and V. T. Parker, eds. 1998. Ecological Scale: Theory and Application. Columbia University Press, New York.

Plafkin, J. L., M. T. Barbour, K. D. Porter, S. K. Gross, and R. M. Hughes. 1989. Rapid Bioassessment Protocols for Use in Streams and Rivers: Benthic Macroinvertebrates and Fish. EPA/440/41-89-001. Assessment and Water Protection Division, U.S. Environmental Protection Agency, Washington, DC.

Platt, T., and K. L. Denman. 1975. Spectral analysis in ecology. Annual Review of Ecology and Systematics 6:189-210.

Poff, N. L., J. D. Allan, M. B. Bain, J. R. Karr, K. L. Prestegaard, B. D. Richter, R. E. Sparks, and J. C. Stromberg. 1997. The natural flow regime: A paradigm for river conservation and restoration. BioScience 47:769-784.

Post, W. M., and J. Pastor. 1996. Linkages - an individual-based forest ecosystem model. Climatic Change 34:253-261.

Potter, C. S., J. T. Randerson, C. B. Field, P. A. Matson, P. M. Vitousek, H. A. Mooney, and S. A. Klooster. 1993. Terrestrial ecosystem production: A process model based on global satellite and surface data. Global Biogeochemical Cycles 7(4):811-841.

Prentice, I. C., M. T. Sykes, and W. Cramer. 1993. A simulation model for the transient effects of climate change on forested landscapes. Ecological modelling 65:51-70.

Preston, F. W. 1962. The canonical distribution of commonness and rarity. Ecology 43:185-215; 410-432.

Preston, F. W. 1960. Time and space and the variation of species. Ecology 41:785-790.

Pulliam, H. R. 1988. Sources, sinks and population regulation. Amer. Natur. 132:652-661.

Pulliam, H. R., and B. J. Danielson. 1991. Sources, sinks, and habitat selection:A landscape perspective on population dynamics. Amer. Natur. 137:S50-S66.

Rabalais, N. N., W. J. Wiseman, Jr., R. E. Turner, D. Justic, B. K. Sen Gupta, and Q. Dortch. 1996. Nutrient changes in the Mississippi River and system responses on the adjacent continental shelf. Estuaries 19:386-407.

Race, M. S., and M. S. Fonseca. 1996. Fixing compensatory mitigation: What will it take? Ecological Applications 6:94-101.

Raich, J. W., E. B. Rastetter, J. M. Melillo, D. W. Kicklighter, P. A. Steudler, B. J. Peterson, A. L. Grace, B. Moore III, and C. J. Vorosmarty. 1991. Potential net primary productivity in South America: Application of a global model. Ecological Applications 1(4):399-429.

Ramsey, F. L., and D. W. Schafer. 1997. The Statistical Sleuth: A Course in Methods of Data Analysis. Duxbury Press, Belmont, CA.

Ratcliffe, D. A. 1967. Decrease in eggshell weight in certain birds of prey. Nature 215:208-210.

Reeves, D.W. 1997. The role of soil organic matter in maintaining soil quality in continuous cropping systems. Soil and Tillage Research 43:131-167.

Renberg, I. 1990. A 12,600 year perspective of the acidification of Lilla Öresjön, southwest Sweden. Philosophical Transactions of the Royal Society of London, B. 327:357-361.

Repetto, R., W. Magrath, M. Wells, C. Beer, and F. Rossini. 1989. Wasting Assets:Natural Resources in the National Income Accounts. World Resources Institute, Washington, D.C.

Reynoldson, T. B., R. H. Norris, V. H. Resh, K. E. Dey, and D. M. Rosenberg. 1997. The reference condition: A comparison of multimetric and multivariate approaches to assess water-quality impairment using benthic macroinvertebrates. Journal of the North American Benthological Society 16(4):833-852.

Riley, M. J., and H. G. Stefan. 1987. A dynamic lake water quality simulation model. Ecol. Model. 43:155-182.

Ripley, B. D. 1981. Spatial Statistics, John Wiley & Sons, New York.

Root, T. L. 1993. Effects of global climate change on North American birds and their communities. Pp. 280-292. in P. Kareiva, J. Kingsolver, and R. Huey, eds., Biotic Interactions and Global Change. Sinauer Associates, Sunderland, MA.

Root, T. L., and L. McDaniel. 1995. Winter population trends of selected songbirds. Pp. 21-23 in E. T. LaRoe, G. S. Farris, C. E. Puckett, P. D. Doran, and M. J. Mac, eds. Our Living Resources: A Report to the Nation on the Distribution, Abundance, and Health of U.S. Plants, Animals, and Ecosystems. U.S. Department of the Interior, National Biological Service, Washington D.C.

Rosenzweig, M. L. 1998. Preston's ergodic conjecture:The accumulation of species in space and time. In M. L. McKinney, ed. Biodiversity Dynamics. Columbia University Press, New York.

Rosenzweig, M. L. 1995. Species Diversity in Space and Time. Cambridge Univ. Press, Cambridge, U.K.

Sado, K., and M. M. Islam. 1996. Effect of land cover on areal evapotranspiration using Landsat TM data with meteorological and height data:The case of Kitami City, Japan. Hydrolog. Sci. J. 41:207.

Sagoff, M. 1996. On the value of endangered species. Environmental Management 20(6):897-911.

Sakamoto, M. 1966. Primary production by the phytoplankton community in some Japanese lakes and its dependence on depth. Arch. Hydrobiol. 62:1-28.

Sarmiento, J., C. LeQuere, and S. Pacala. 1995. Limiting future atmospheric carbon dioxide. Glob. Biogeochem. Cycles 9:121-137.

Sauer, J. R., J. E. Hines, G. Gough, I. Thomas, and B. G. Peterjohn. 1997. The North American Breeding Bird Survey Results and Analysis, Version 96.4. Patuxent Wildlife Research Center, Laurel, Md.

Schlesinger, W. H. 1997. Biogeochemistry: An Analysis of Global Change. 2nd. ed. Academic Press, New York.

Schnoor, J. L. 1981. Fate and transport of dieldrin in Coralville Reservoir: Residues in fish and water following a pesticide ban. Science 211:840-842.

Schumann, G. L. 1991. Plant Diseases and Their Social Impact. The American Phytopathological Society. St. Paul, MN.

Sellers, P. J., R. E. Dickinson, D. A. Randall, A. K. Betts, F. G. Hall, J. A. Berry, G. J. Collatz, A. S. Denning, H. A. Mooney, C. A. Nobre, N. Sato, C. B. Field, and A. Henderson-Sellers. 1997. Modeling the exchange of energy, water, and carbon between continents and the atmosphere. Science 275:502-509.

Shannon, C. E., and W. Weaver. 1964. The Mathematical Theory of Communication. University of Illinois Press, Urbana.

Sharpley, A., and M. Meyer. 1994. Minimizing agricultural non-point-source impacts:A symposium overview. J. Environ. Qual. 23:1-3.

Sharpley, A. N., J. J. Meisinger, A. Breeuwsma, T. Sims, T. C. Daniel, and J. S. Schepers. 1996. Impacts of animal manure management on ground and surface water quality. Pp. 1-50 in J. Hatfield, ed. Effective Management of Animal Waste as a Soil Resource. Lewis Publ. Boca Raton, FL.

Shmida, A., and S. Ellner. 1984. Coexistence of plant species with similar niches. Vegetatio 58:29-55.

Shugart, H. H. 1984. A Theory of Forest Dynamics:The Ecological Implications of Forest Succession Models. Springer-Verlag, Berlin.

Silver, W. L., T. G. Siccama, C. Johnson, and A. H. Johnson. 1991. Changes in red spruce populations in montane forests of the Appalachians, 1982-1987. Am. Midland Naturalist 125:340-347.

Simberloff, D. 1990. Hypotheses, errors, and statistical assumptions. Herpetologica 46:351-357.

Simenstad, C. A., J. A. Estes, and K. W. Kenyon. 1978. Aleuts, sea otters, and alternate stable-state communities. Science 200:403-411.

Sims, J. T., and D. C. Wolf. 1994. Poultry waste management: Agricultural and environmental issues. Advances in Agronomy 52:1-83.

Sims, J. T., R. R. Simard, and B. C. Joern. 1998. Phosphorus loss in agricultural drainage: Historical perspective and current research. J. Environ. Qual. 27:277-293.

Simpson, R. S., and R. C. Houts. 1971. Fundamentals of Analog and Digital Communication Systems. Allyn and Bacon, Boston, MA.

Smith, R. A., G. E. Schwarz, and R. B. Alexander. 1997. Regional interpretation of water-quality monitoring data. Water Resources Research 33(12):2781-2798.

Skole, D., and C. J. Tucker. 1993. Tropical deforestation and hibitat fragmentation in the Amazon:Satellite data from 1978 to 1988. Science 260:1905-1910.

Skole, D., C. O. Justice, A. Janetos, and J. R. G. Townshend. 1997. A Land Cover Change Monitoring Program:A Strategy for International Effort. Mitigation and Adaptation Strategies for Global Change. Kluwer Academic Publishers, Dordrecht, The Netherlands.

Soil Survey Staff. 1993. Soil Survey Manual. USDA Handbook 18. Government Printing Office, Washington, DC.

Solomon, A. M. 1986. Transient responses of forests to CO_2-induced climate change:Simulation modelling experiments in eastern North America. Oecologia 68:567-579.

Solomon, A. M., and H. H. Shugart. 1984. Integrating forest stand simulations with paleoecological records to examine long-term forest dynamics. Pp. 333-356 in G. I. Ågren, ed. State and Change of Forest Ecosystems—Indicators in Current Research. Swedish University of Agricultural Sciences, Report Number 13, Uppsala, Sweden.

Spaulding, R. F., and M. E. Exner. 1993. Occurrence of nitrate in groundwater:A review. J. Environ. Qual. 22:392-402.

Spies, T. A., J. F. Franklin, and T. B. Thomas. 1988. Coarse woody debris in Douglas-fir forests of western Oregon and Washington. Ecology 69:1689-1702.

Stefan, H. G., M. Hondzo, X. Fang, J. G. Eaton, and J. H. McCormick. 1995. Predicted effects of global climate changes on fishes in Minnesota lakes. Pp. 57-72 in R. J. Beamish, ed. Climate Change and Northern Fish Populations. Can. Spec. Publ. Fish. Aquat. Sci. 121. National Research Council Canada, Ottawa.

Stefan, H. G., M. Hondzo, X. Fang, J. G. Eaton, and J. H. McCormick. 1996. Simulated long-term temperature and dissolved oxygen characteristics of lakes in the north-central United States and associated fish habitat limits. Limnol. Oceanogr. 41:1124-1135.

Strauss, D. J. 1990. Have There Been Growth Changes in the Sierra Nevada: A Statistical Analysis. Technical Report No. 181. University of California, Riverside.

Streeter, H. W., and E. B. Phelps. 1925. A study of the pollution and natural purification of the Ohio River. III. Factors concerned in the phenomena of oxidation and reaeration. U.S. Public Health Service, Public Health Bull. Washington, D.C.

Sudhakar, S., and D. K. Pal. 1993. Water quality assessment of Lake Chilka. Int. J. Remote Sens. 14:2575.

Susskind, L. E., and L. Dunlap. 1981. The importance of nonobjective judgments in environmental impact assessments. Environmental Impact Assessment Review 2:335-366.

Suter, G. W., II. 1993. Ecological Risk Assessment. Lewis Publishers, Boca Raton, FL.

Swengel, A. B. 1995. Fourth of July butterfly count. Pp. 171-172 in E. T. LaRoe, G. S. Farris, C. E. Puckett, P. D. Doran, and M. J. Mac, eds. Our Living Resources: A Report to the Nation on the Distribution, Abundance, and Health of U.S. Plants, Animals, and Ecosystems. U.S. Department of the Interior, National Biological Service, Washington DC.

Swetnam, T. W., and J. L. Betancourt. 1998. Mesoscale disturbance and ecological response to decadal scale climatic variability in the American Southwest. Journal of Climate 11:3128-3147.

Thomann, R. V., and J. A. Mueller. 1987. Principles of Surface Water Quality Modeling and Control. Harper & Row, New York.

Tilman, D. 1996. Biodiversity: Population versus ecosystem stability. Ecology 77:350-363.

Tilman, D., and J. A. Downing. 1994. Biodiversity and stability in grasslands. Nature 367:363-365.

Townshend, J. R. G., C. O. Justice, D. Skole, J. P. Malingreau, J. Cihlar, P. Teillet, F. Sadowski, and S. Ruttenberg. 1994. The l km resolution global data set:Needs of the International Geosphere Biosphere Programme. International Journal of Remote Sensing 15:3417-3441.

Trautman, M. B. 1981. The Fishes of Ohio. Second Edition. Ohio State University Press, Columbus.

Turner, B. I., II, R. H. Moss, and D. L. Skole. 1993. Relating land-use and global land-cover change. IGBP Report 24. IGBP Secretariat, Stockholm.

U.S. Bureau of Census. 1987. 1987 Census of Agriculture, Final County File. Census Bureau, Washington, D.C.

USDA. 1997. Agricultural Resources and Environmental Indicators, 1996-97. USDA Economic Research Service, Natural Resources and Environment Division. Agricultural Handbook Number 712.

USDA Forest Service. 1994. Forest Health Monitoring: A National Strategic Plan. USDA Forest Service, Southern Region, Atlanta, Ga.

U.S. Environmental Protection Agency (U.S. EPA). 1990. County-level fertilizer sales data. Rep PM-221. Office of Policy Planning and Evaluation, Washington, D.C.

U.S. Environmental Protection Agency (U.S. EPA). 1998. National Water Quality Inventory: 1996 Report to Congress. EPA 841-F-97-003, Washington, D.C.

U.S. Soil Conservation Service. 1992. Agricultural waste management field handbook. Chapter 4 in National Engineering Handbook. U.S. Natural Resources Conservation Service, Washington, D.C.

Usher, M. B. 1981. Modelling ecological succession, with particular reference to Markovian models. Vegetatio 46:11-18.

Van der Molen, D. T., A. Breeuwsma, and P. C. M. Boers. 1998. Agricultural nutrient losses to surface water in the Netherlands: Impact, strategies and perspectives. J. Environ. Qual. 27:4-11.

van der Ploeg, R. R., H. Ringe, G. Machulla, and D. Hermsmeyer. 1997. Postwar nitrogen use efficiency in West German agriculture and groundwater quality. J. Environ. Quality 26:1203-1212.

Van Hulst, R. 1979. On the dynamics of vegetation:Markov chains as models of succession. Vegetatio 40:3-14.

Vannote, R. L., G. W. Minshall, K. W. Cummins, J. R. Sedell, and C. E. Cushing. 1980. The river continuum concept. Canad. J. Fish. Aquat. Sci. 37:130-137.

VEMAP Member. 1995. Vegetation/ecosystem modeling and analysis project. Comparing biogeography and biogeochemistry models in a continental-scale study of terrestrial ecosystem responses to climate change and CO_2 doubling. Global Biogeochemical Cycles 9(4):407-434.

Vitousek, P., H. A. Mooney, J. Lubchenco, and J. M. Melillo. 1997. Human domination of Earth's ecosystems. Science 277:494-499.

Vogelmann, J. E. 1988. Detection of forest change in the Green Mountains of Vermont using multispectral scanner data. International Journal of Remote Sensing 9:1187-1200.

Vogelmann, J. E., and B. N. Rock. 1989. Assessing forest damage in high-elevation coniferous forests in Vermont and New Hampshire using Thematic Mapper data. Remote Sensing of the Environment 24:227-246.

Vogelmann, J. E., T. Sohl, and S. M. Howard. 1998. Regional characterization of land cover using multiple sources of data. Photogram. Engrg. Remote Sens. 64:45.

Vollenweider, R. A. 1976. Advances in defining critical loading levels for phosphorus in lake eutrophication. Mem. Ist. Ital. Idrobiol. 33:53-83.

Vollenweider, R. A. 1975. Input-output models with special reference to the phosphorus loading concept in limnology. Schweiz. A. Hydrol. 37-53-84.

Vollenweider, R. A. 1969. Möglichkeiten und Grenzen elementarer Modelle der Stoffbilanz von Seen. Arch. Hydrobiol. 66:1-36.

Walker, W. W. 1987. Empirical Methods for Predicting Eutrophication in Impoundments. Report 4, Phase III: Applications Manual. Tech. Report. E-8-9, U.S. Army Engineer Experiment Station, Vicksburg, MS.

Waring, R. H. 1982. Estimating forest growth and efficiency in relation to canopy leaf area. Advances in Ecological Research 13:327-354.

Washington, H. G. 1984. Diversity, biotic, and similarity indices: A review with special relevance to aquatic ecosystems. Water Research 18:153-694.

Wessman, C. A., J. D. Aber, D. L. Peterson, and J. M. Melillo. 1988. Remote sensing of canopy chemistry and nitrogen cycling in temperate forest ecosystems. Nature 335: 154-156.

Whitaker, R. H., and G. E. Likens. 1973. Carbon in the biota. Pp. 281-302 in G. M. Woodwell, and E. V. Pecan, eds. Carbon and the Biosphere, Washington, DC.

Whittier, T. R., and E. T. Rankin. 1992. Regional patterns in three biological indicators of stream condition in Ohio. Pp. 975-995 in D. H. McKenzie, D. E. Hyatt, and V. J. McDonald, eds. Biological Indicators. Volume 2. Elsevier Applied Science, London and New York.

Wickerhauser, M. V. 1994. Adapted Wavelet Analysis from Theory to Software. A.K. Peters, Wellesley, Massachusetts.

Williamson, C. B. 1943. Area and the number of species. Nature 152:264-267.

Wischmeier, W. H., and D. D. Smith. 1978. Predicting Rainfall Erosion Losses – A Guide To Conservation Planning. Agriculture Handbook 537. Science and Education Administration, U.S. Department of Agriculture, Washington, D.C.

Wofsy, S. C., M. L. Goulden, J. W. Merger, S.-M. Fass, P. S. Bakwin, B. C. Daube, S. L. Banow, and F. A. Bazzaz. 1993. Net exchange of CO_2 in a unit-latitude forest. Science 260:1314-1317.

Wood, E. F., M. Sivapalan, and K. Beven. 1990. Similarity and scale in catchment storm response. Reviews of Geophysics 28:1-18.

Woodwell, G. M., F. T. Mackenzie, R. A. Houghton, M. J. Apps, E. Gorham, and E. A. Davidson. 1995. Will the warming speed the warming? Pp. 393-411 in Biotic Feedbacks in the Global Climate System. G. M. Woodwell and F. T. Mackenzie, eds. Oxford University Press, New York.

Wright, Jr., H. E. 1971. Late Quaternary vegetational history of North America. Pp. 425-464 in The Late Cenozoic Glacial Ages. K. K. Turekian, ed. Yale University Press, New Haven, CT.

Wydoski, R. S., and R. R. Whitney. 1979. Inland Fishes of Washington. University of Washington Press, Seattle.

Young, R. A., C. A. Onstad, D. D. Bosch, and W. P. Anderson. 1989. AGNPS:A nonpoint-source pollution model for evaluating agricultural watersheds. Journal of Soil and Water Conservation 44(2):168-172.

Appendixes

Appendix A

Variability, Complexity, and the Design of Sampling Procedures

No general laws exist that allow us to predict the relative magnitude of temporal and spatial variability of different types of parameters across the diversity of ecological systems. Variability is highly dependent on the temporal and spatial scales of the data sets and on the level of aggregation of the parameter of interest (e.g., species level versus community level [Allen and Starr 1982, Frost et al. 1988, Wood et al. 1990]). This scale-dependence raises a complication in comparative studies because it is possible to confound differences in patterns of variability with differences in scales of measurement made at two or more systems.

To compute a nationwide indicator, a variety of different data must be aggregated, and the choice of aggregation level may be one of the most important decisions in the design of an indicator and of the monitoring program that generates the necessary input data. For biological data, levels of aggregation (e.g., species, guilds, major groups) has greater effect on observed variability than does spatial or temporal extent of the data (Kratz 1995).

Parameters useful in monitoring programs designed to detect trends and patterns often have two, potentially conflicting characteristics. On the one hand, they must be sufficiently sensitive to environmental conditions to indicate changes that occur. On the other hand they must not exhibit so much natural variability as to mask detection of significant changes in environmental conditions. Thus, understanding the relative variability and sensitivity of parameters is important in choosing optimal parameters for a monitoring program.

In addition, use of data from long-term monitoring programs is often made difficult because such monitoring programs are typically not designed as experiments. Monitoring is usually conducted to assess the effects of certain agents, procedures, or programs that cannot be subjected to controlled experimentation (Cochran 1983). Cause-and-effect relationships usually cannot be inferred from such data because confounding variables (identified or not) that were not controlled during the study may be responsible for the observed differences (Ramsay and Schafer 1997). Nonetheless, there are approaches that can reduce the likelihood of drawing incorrect conclusions from long-term data sets (Arbaugh and Bednar 1996). These approaches are characterized by:

- comparison of "quasi-treatment" to "quasi-control" groups to resemble a designed experiment;
- comparison of "treatment" groups with more than one "control" group to develop different contrasts with the "treatment" group;
- comparison of "treatment" and "control" groups with important exogenous variables; if this is unfeasible, groups should be adjusted for differences using covariates in the analysis; and
- use of a variety of measures/components to reduce the dependence of study results on a single aspect of the data and on assumptions inherent for single methods of analysis (Strauss 1990).

It is impossible to optimize both experimental and monitoring goals simultaneously, but long-term monitoring data can be used as key inputs in the development of estimates of the states of the systems being monitored, together with confidence levels associated with those estimates. However, careful consideration must be given to statistical treatment of correlated observations when variance estimates and confidence intervals are calculated (Conquest 1993).

TEMPORAL BEHAVIOR OF ECOSYSTEMS AND SAMPLING DESIGN

The natural world oscillates. Such oscillations are well known to ecologists and include many classic population cycles such as the hare-lynx and lemming cycles (Elton 1942, Keith 1963). The frequency and amplitude of such oscillations, which are characteristic properties of any complex system, pose problems in sampling to detect trends in a particular property. For example, suppose that an annual sampling strategy over ten years has revealed a steady decline (or increase) in productivity. Such a ten-year record is relatively long for most monitoring programs. Can we say with confidence that the system is deteriorating (or improving) as a result of worse (or better) growing conditions? We cannot unless we can separate such a trend from normal oscillations: we may simply have caught the system in the descending (or ascending) portion of a 30-year oscillation whose mean is stationary.

To detect trends, we must sample at a frequency from which we can characterize the oscillatory character of systems. However, to determine the natural frequencies from real data, we must sample for a long time. For example, to determine whether a system has a 30-year oscillation, we must sample for 60 years to obtain data on at least two repeated oscillations. In many cases, by the time we have a sufficiently long time series, the monitoring strategy will simply confirm what may already be obvious. Conversely, we may have spent much time and effort (and money) sampling the system too frequently to detect a relatively long cycle. For example, it would make no sense whatsoever to sample at daily frequencies to detect trends embedded within a 30-year cycle; even annual or biannual sampling would be too frequent.

These considerations pose three problems: identifying reasonable long-term surrogates for current ecosystems to serve as first approximations to characterize oscillatory behavior, determining the optimal sampling frequency of an ecosystem property from these surrogates, and detecting changes in the behavior of an oscillating system.

Simulation models and long-term paleorecords can be used as first approximation surrogates of the real system to obtain a long time series about attributes of interest (species composition, biomass, productivity, soil nitrogen, etc.) under changing conditions for which we desire to make policy (climate warming, etc.).

The advantage of paleorecords is that they provide real data on the historical behavior of particular ecosystem properties at a given location. Traditionally, pollen analyses have been the main tool of paleoecologists, yielding data on changes in species composition surrounding a catchment basin. However, recent analyses include correlating changes in vegetation with charcoal abundance (Clark 1988a and b, Clark et al. 1989) and biogeochemical analyses of sediment (MacDonald et al. 1993). A particularly useful paleorecord can be obtained from cores taken from varved lakes. A varved lake sediment is one in which annual cycles of deposition are clearly visible in the sediment column, enabling paleoecologists to reconstruct an annual time series of changes in species composition and associated biogeochemical properties and fire regimes (e.g., Clark 1988a and b). In the absence of varved sediments, fine-increment sampling and description of layers also enables a paleoecologist to reconstruct a detailed time series, particularly when layers are radiocarbon dated to anchor the chronology (MacDonald et al. 1993, Cooper and Brush 1991). Tree-ring records are another source of data on the historical behavior of multiple, often interacting, ecosystem properties. For example, tree rings provide multi-century to multi-millennial histories of seasonally resolved climate as well as independent records of ecological responses to climate such as changes in fire regime and forest-stand structure and composition (Swetman and Betancourt 1998, Lloyd and Graumlich 1997).

Limits of the paleorecord are that only certain types of data are available, and the causes of temporal patterns must be inferred from the record itself or from ancillary data. Simulation models as surrogates of real systems may allow the researcher to overcome these problems, but the simulation models themselves must be extensively tested first. Even then, the model behavior is often extended into domains beyond that for which the model was originally intended or parameterized. Nonetheless, simulation models are useful tools—in many cases the best available tool—that the researcher can use to generate a long time series of a particular ecosystem property and examine its expected behavior under different stressors, such as climate change (Shugart 1984; Solomon 1986; Pastor and Post 1988, 1993; Ågren et al. 1991; Cohen and Pastor 1991; Post and Pastor 1996; Prentice et al. 1993). Such simulation models can help test hypotheses regarding causes of changes seen in the paleorecord (Solomon and Shugart 1984).

To determine the optimal sampling frequency and to examine how the system responds to rapid changes, the surrogate time series of data can then be analyzed using techniques of signal processing (Shannon and Weaver 1964, Simpson and Houts 1971, Wickerhauser 1994). Fourier analysis and wavelet analysis are the two main techniques of interest.

A system may oscillate with many different frequencies, each with different amplitudes. Fourier analysis—sometimes called spectral analysis—decomposes this complex behavior into a sum of sine and cosine waves of specific frequencies and amplitudes (Shannon and Weaver 1964, Simpson and Houts 1971, Wickerhauser 1994). A particularly good review of the application of this techniques to ecological problems, including the paleorecord and output from simulation models, is given by Platt and Denman (1975). One objective of a Fourier decomposition is to identify the frequencies with the greatest amplitude because these are the frequencies that most strongly govern the system behavior. To accomplish this, frequencies of the sine and cosine waves are plotted against their amplitude. These plots are often called spectral plots.

Once a spectral plot is constructed, a powerful sampling theorem—the Nyquist theorem—can be used to specify the optimal sampling frequency. This theorem states that the time between samples of an oscillatory signal need be no greater than half the frequency of the shortest cycle. This frequency is known as the Nyquist frequency. Sampling more frequently than the Nyquist frequency does not yield any additional information and simply adds to the cost of the sampling program. By first determining the Nyquist sampling frequencies for different properties in the surrogate data sets, the researcher can determine the optimal time that current sites need to be resampled to characterize their behavior.

Although Fourier analysis determines the relative contributions of different oscillatory frequencies to system behavior, it is less useful for determining when sudden changes in frequencies occur. These sudden changes may be early warning signs of changes in the status of a system. For example, both simulation models (Solomon 1986, Pastor and Post 1988) and the paleorecord (MacDonald et al. 1993) have often shown that productivity and species composition of forests change rapidly in response to gradual changes in climate once threshold temperatures for particular species are exceeded.

To detect these sudden changes, the related and newer tool of wavelet analysis is useful (Wickerhauser 1994, Hubbard 1996). Instead of determining the contribution of each frequency to the entire time series signal as Fourier analysis does, wavelet analysis determines when there are sudden changes in frequency, times of rapid change in system behavior. This would then alert the researcher to investigate further whether this change is a normal aspect of system functioning or whether it is imposed by changes in a stressor. Applying wavelet analysis to surrogate data sets with known behavior to changes in a stressor would help us recognize the "symptoms" of sudden change in real word systems. Because of the recent development of this mathematical technique, the application of wavelet analysis to ecological problems (e.g., Bradshaw and McIntosh 1993, Bradshaw and Spies 1992) would benefit from a focused research program. If the wavelet behavior of paleorecords or simulation models can be characterized with respect to stressors such as global warming, then we would have a set of likely "symptoms" to alert policy makers to impending problems.

AN EXAMPLE OF FOURIER ANALYSIS OF SIMULATION MODEL OUTPUT TO DETERMINE OPTIMAL SAMPLING FREQUENCIES

We present here an example of using a simulation model to generate a long time series of output as a first approximation of the behavior of a forested ecosystem. The output is then analyzed using Fourier analysis to determine the optimal frequency to sample an oscillating system under current climates.

Many ecosystems are now well described by extensively tested simulation models that incorporate fundamental biological processes respon-

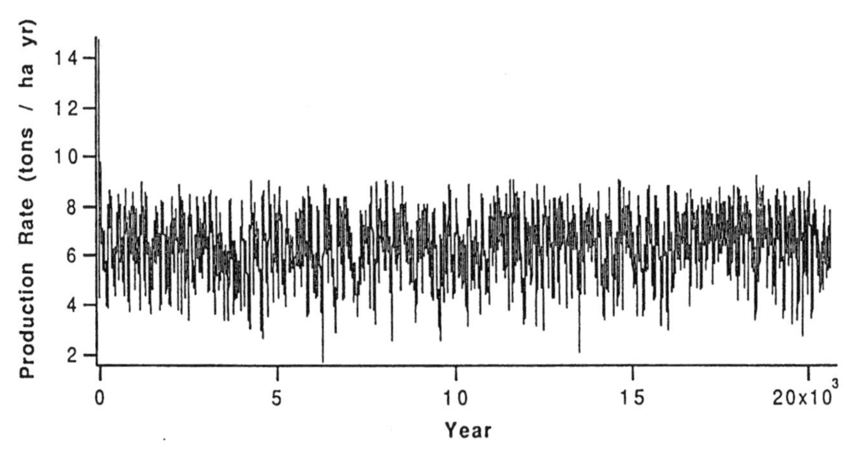

FIGURE A-1. Time series of model output of productivity of a forest in northeastern Minnesota on a silty clay loam soil under current climate. Model runs were made with LINKAGES. Adapted from Pastor and Cohen 1997.

sible for system oscillations. Among the best known are individual-based models of forest ecosystems, known collectively as JABOWA-FORET models (Shugart 1984). Emanuel et al. (1978) provide an excellent example of how to apply Fourier analysis to the output from such models to determine optimal sampling rates in the field. We use this approach here to analyze cycles of productivity from a forest ecosystem model under current climate for northeastern Minnesota (Pastor and Cohen 1997), from which we calculate the optimal sampling rate to characterize these cycles.

We first generate a long time series of productivity under current climate for northern Minnesota for a single plot of 0.01 ha on a silty clay loam soil with high water holding capacity (Figure A-1). Such a plot represents a standard forest inventory plot that would be resampled in a monitoring program such as the USFS Continuous Forest Inventory and Analysis (CFIA) System, which supplies the raw data for timber policy in this country.

We then use Fourier analysis to decompose this output stream or signal into a sum of sine and cosine functions each of characteristic frequency and amplitude according to:

$$x(t) = \frac{1}{2\pi} \int_{-\infty}^{\infty} \mathrm{A}(\omega) e^{(i2\pi\omega t)} dt$$

where $x(t)$ is the data or signal at time t (Figure A-1), $A(\omega)$ is a coefficient (often called a Fourier coefficient) that specifies the power or amplitude of a cycle of frequency ω, and $i = \sqrt{-1}$. The coefficients and frequencies are determined such that each exponential function and its associated coefficient is orthogonal to the others. The oscillatory nature of the above equation can be clarified by recalling that

$$e^{i2\pi\omega t} = \cos(2\pi\omega t) + i\sin(2\pi\omega t).$$

We essentially thereby obtain a series of cosine functions each of characteristic frequency and amplitude that describe the time series of productivity under northeastern Minnesota climate. The next step is to plot the amplitude against the frequency of the data to obtain graphs, known as spectral density plots (Figure A-2).

We are now ready to apply the Nyquist sampling theorem to this spectral density plot. This theorem (Nyquist 1928) states that a signal (i.e., Figure A-1) whose highest frequency occurs at B cycles per unit time (i.e., from Figure A-2) is uniquely determined by sampling at a uniform rate of

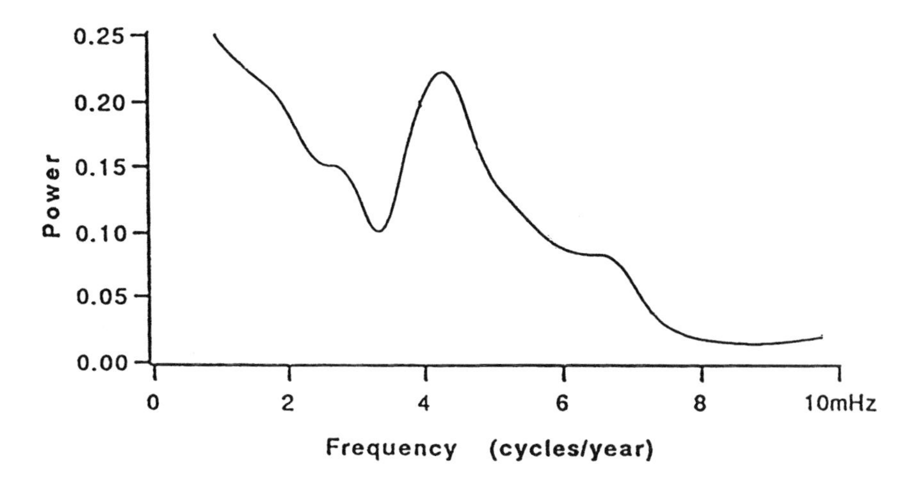

FIGURE A-2 Spectral analysis of the time series of productivity in Fig. A-1. Note the peak frequency at approximately 0.043 cycles per year (1 cycle per year = 1 Hz; x-axis in Hz/100), which corresponds to a period of 23 years. One half this period, or 11-12 years, is the optimal sampling frequency according to the Nyquist sampling theorem. Adapted from Pastor and Cohen 1997.

at least 2B samples per unit time. In other words, the time between samples need be no greater than half the frequency of the shortest cycle, known as the Nyquist frequency.

For the above plot, the highest frequency occurs at approximately 0.043 cycles per year. The inverse of the frequency is the period, or 23 years. This period is a result of the dynamics of canopy death and replacement. One half of this, or 11-12 years, is the Nyquist sampling frequency. Therefore, to characterize trends in forest productivity in a monitoring plot, one need sample no more often than 11-12 years. Interestingly, the CFIA plots are sampled on average once every 10 years, and Emanuel et al. (1978) found a Nyquist sampling frequency of 10 years for an Appalachian hardwood forest using an earlier version of this model. Therefore, the current sampling frequencies of the CFIA and Forest Health Monitoring Programs appear to sample forests sufficiently often to characterize productivity, at least in northern Minnesota on clay soils.

The above model run was for a silty clay loam soil. Similar runs could be made for other soil types, such as sands. One might find that other soils require different sampling frequencies to characterize productivity. One might also apply the technique to other model output of interest (soil nitrogen, biomass, foliage height diversity, species diversity, etc.) to determine optimal sampling rates for other properties. Similar approaches could be used to determine changes in sampling frequency under different climates, with increasing or decreasing acidity of precipitation, with different fertilizer application levels or other parameters of interest.

Appendix B

Markov Matrices of Landscape Change

Developments in geographic information systems and remote sensory image analysis have made it possible to calculate changes in land cover classes during selected time intervals. The data are assembled in matrices, often known as *change matrices,* and the analyses are called *landscape-change analyses* (Vogelmann 1988, Vogelmann and Rock 1989, Lozano-Garcia and Hoffer 1985). Such analyses are valuable because they tell us what has happened over a region during some time interval in the past. However, by their nature they are retrospective. In contrast, policy must be based on predictive analyses of landscape trajectories given the current rates of changes from one land type to the others. The theory of Markov chains provides the mathematical basis for at least a first approximation of the consequences of current trends in land cover distributions.

This theory encompasses a large body of literature, most recently reviewed by Baker (1989) and Pastor et al. (1992). A Markov chain consist of a vector x of the distribution of land covers at time t and a matrix $A(\tau)$ of transition probabilities of changes from each land cover class to the others during a time period (τ):

$$x_{t+\tau} = A(\tau)x_t.$$

To parameterize a Markov chain of landscape dynamics, a map of the landscape at time t is subdivided into pixels which are assigned individually into one of m classes. Classes can be assigned to each pixel taxonomically (that is, the pixel is occupied by a particular species; Horn 1975, Lippe et al. 1985), through the use of multivariate cluster or principal components analyses (Van Hulst 1979, Usher 1981), or through remotely sensed data such as air photo analyses (Johnston and Naiman 1990, Pastor et al. 1992) or satellite imagery (Hall et al. 1991). To obtain transition probabilities, a second map is then prepared for time $t + \tau$. The two maps are overlaid atop one another and the number of pixels that changed during τ units of time from one land cover to another are then enumerated. The maximum likelihood estimates of probabilities of change from one land-cover to another during time interval τ are:

$$p_{i,j,\tau} = \frac{n_{i,j}}{\sum_{j=1}^{m} n_{i,j}},$$

where $p_{i,j,t}$ are the transition probabilities from land cover i to land-cover j in time interval t, and $n_{i,j}$ are the number of such transitions across all pixels of the landscape of m land cover classes.

When the time interval of the model (i.e., annual or decadal) is something other than the desired time interval of the two maps (as frequently happens when using a historic set of air photos), then the probabilities of change can be normalized to the desired time step (Pastor et al. 1992) as follows:

$$\begin{cases} p_{i,j} = 1 - e^{(\ln(1-p_{i,j,\tau}))/\tau} & \text{when } i \neq j \\ p_{i,j} = 1 - \sum_{j=1}^{n} p_{i,j} & \text{when } i = j \end{cases}$$

where τ is expressed as some fraction of the desired time scale. For example, if transition probabilities are calculated from data layers taken 13 years apart and the user wishes transition probabilities to be expressed in decadal increments, then $\tau = 1.3$ in the equation above.

We are now in a position to use the matrix of transition probabilities to guide policy. Suppose a particular policy is formulated to move the landscape from the current land cover vector to some desired future state. The policy is implemented for, say, ten years. A new map of land cover distribution is made from the monitoring data after 10 years and a matrix of transition probabilities is calculated as above. The question is: Is the landscape headed toward the desired distribution of land cover classes and, if so, how long will it take to get there under the new policy?

Two properties of the matrix, known as the eigenvalues and eigenvectors, are useful for answering these questions. These satisfy the equation:

$$A\mu = \lambda\mu$$

where A is the matrix of transition probabilities, μ is an eigenvector, and λ is an eigenvalue (a scalar). Usually a number of eigenvalues and associated eigenvectors satisfy this equation; these are easily calculated with current software packages.

If all of the transition probabilities are greater than zero, then any one land cover class can be reached from any other. The matrix is then said to be irreducible (Caswell 1989, Pastor et al. 1992). Because all columns in an irreducible Markov matrix sum to 1, the dominant (largest) eigenvalue equals 1. The eigenvector of the distribution of land cover classes associated with the dominant eigenvalue then represents the steady state condition of the landscape. When the land cover vector is in this condition, all the inputs to a land cover class by transition from others equals all the outputs from that land-cover class to all others.

If the matrix is not irreducible (i.e., some transition probabilities equal zero), then the dominant eigenvector is still the steady state distribution of land cover classes if the dominant eigenvalue of the entire matrix equals that of the largest irreducible submatrix (i.e., a submatrix of non-zero transitions probabilities among a subset of land cover classes).

The dominant eigenvector is therefore where the landscape will end up if the current policy is pursued. One can then ask, Is this the desired future condition of the landscape? If not, then policies need to be adjusted. Various alternatives can be determined by "experimenting" with the transition probabilities of the current Markov matrix to see if they yield a new matrix with a dominant eigenvector that matches the desired future conditions.

If the dominant eigenvector does represent the desired future condition of the landscape, then one may ask, How long will it take to get there? To determine this, one must calculate the ratio of the dominant eigenvalue to the absolute value of the second largest eigenvalue. This ratio is known as the damping ratio (Usher 1981, Caswell 1989). The greater this ratio, the faster the approach to steady state.

The approach is exponentially asymptotic and its rate, r, at any given time, t, is

$$r = ke^{-\tau \ln \rho}$$

where ρ is the damping ratio and k is a constant (Caswell 1989). Because the approach to steady state is asymptotic, it is more convenient to calculate the time for some proportion of convergence to steady state, say 95% convergence. This time, t_x, is given by

$$t_x = \ln(x) / \ln(\rho).$$

The percentage of convergence to steady state equals 100 - (100/x). For example, the time required for 95% convergence to steady state is equivalent to the solution of the equation above for $x = 20$ (i.e., 100–(100/20) = 95).

One can now ask not only whether the desired policy is moving the landscape towards the desired future condition, but is it moving it at an acceptable rate. Again, various alternatives to move the landscape faster (or slower) can be determined by "experimentally" changing certain transition probabilities to correspond to alternative policies.

Markov chains lend themselves to hierarchical classification systems. Suppose at the highest level of a classification system there are four land cover classes (say, forests, wetlands, agricultural lands, and urban lands). A simple example of a transition matrix among these land cover classes is given in Table 1a. Most, if not all, of the transition probabilities at such an aggregated level are greater than 0, although they may be very small. That is, usually all transitions occur.

TABLE 1a Matrix Among Four Land Cover Classes

		From: 1	2	3	4
To:	1	X	X	X	X
	2	X	X	X	X
	3	X	X	X	X
	4	X	X	X	X

Now suppose that at the next lower level of the classification, class 1 has 3 subclasses, class 2 has 2 subclasses, class 3 has 3 subclasses, and class 4 has 2 subclasses. A new transition matrix can be calculated for this level of the hierarchy (Table 1b). At this level, it often happens that many transitions do not occur—the transition probability from one land cover class to another is often zero. Such a matrix is known as a "sparse" matrix, and may pose problems for calculations of eigenvectors and eigenvalues unless certain conditions are met (see Caswell 1989 for discussion of this).

However, some interesting properties often emerge. One is that there may be a few land cover classes with many positive transition probabilities through them. In Table 1b, these are land cover classes 1a, 2a, 3b, and 4b. These are particular land cover subclasses through which transitions between the higher level classes commonly take place. It is particularly important to be able to identify and protect these land cover classes. They are analogous to the concept of "keystone species" in community ecology because they control the dynamics of the landscape. In keeping with this analogy, they may be termed "keystone land cover types." Should they be lost because of some land use practice, then transitions between the higher level classes may not happen. These higher level categories may then become decoupled from one another. This decoupling could then preclude the implementation of certain policies that seek to move the landscape into various desired future conditions: it may no longer be possible to achieve the desired future condition because the key land

TABLE 1b Transition Matrix (as Above) but with Subclasses Added

		From:									
		1a	1b	1c	2a	2b	3a	3b	3c	4a	4b
	1a	X	X	X	X	X	X	X	X	X	X
	1B	X	X	X							
	1C	X	X	X							
	2a	X	X	X	X	X	X	X	X	X	X
To:	2b				X	X					
	3a						X	X	X		
	3b	X	X	X	X	X	X	X	X	X	X
	3c						X	X	X		
	4a									X	X
	4b	X	X	X	X	X	X	X	X	X	X

cover class that allows the required transitions may no longer be in existence.

It is obvious that Markov chains are first-order linear models of changes in land cover classes. They are first order because the changes involve no time delays longer than a single time step, and they are linear because the amount of land transferred from one class to another during a time step is simply a portion of the area of each land cover type. However, landscape dynamics are almost certainly nonlinear and often involve time delays. Time delays can be incorporated into Markov chains by extending them to be second order or higher, but the mathematics becomes more complicated. Nonetheless, the theory of higher-order Markov chains (including time delays) and some preliminary applications to species and landscape dynamics have been established (Baker 1989, Acevedo et al. 1995, Kenkel 1993). The application of higher-order Markovian models to behavior of indicators of landscape change would greatly benefit from additional research.

Appendix C

Biographical Sketches of Committee Members and Staff

Gordon H. Orians is professor emeritus of zoology at the University of Washington in Seattle. He received a Ph.D. in zoology in 1960 from the University of California, Berkeley. From 1976 to 1986 he was director of the Institute of Environmental Studies at the University of Washington. Dr. Orians' research interests are evolution of vertebrate social systems; factors determining the number of species an environment will support on a sustained basis; plant-herbivore interactions; and ecology of rare species. He is a member of the National Academy of Sciences.

Martin Alexander is Liberty Hyde Bailey Professor of Soil Science at Cornell University. He has chaired or been a member of a variety of advisory committees to the U.S. Environmental Protection Agency, National Research Council, National Institutes of Health, and U.S. Army, Food and Agricultural Organization of the United Nations and UNESCO, and has consulted with many private companies on environmental pollution. Dr. Alexander's research interests are in the areas of soil and environmental microbiology, and microbial transformations that are of environmental or agricultural importance in natural environments. He received a B.S. in 1951 from Rutgers University, and an M.S., in 1953, and Ph.D., in 1955, from the University of Wisconsin.

Patrick L. Brezonik is professor of environmental engineering at the University of Minnesota and director of the university's Water Resources Research Center. His research interests are the eutrophication of lakes, nitrogen dynamics in natural water, acid rain, and organic matter in water. He is a member of the Water Science and Technology Board and served on the National Research Council's Committee on Inland Aquatic Ecosystems and Committee to Review the Environmental Protection Agency's Environmental Monitoring and Assessment Program. Previously, he was professor of water chemistry and environmental science at the University of Florida. He obtained a Ph.D. in water chemistry in 1968 from the University of Wisconsin.

Grace Brush is a professor in the Department of Geography and Environmental Engineering at the Johns Hopkins University. She received a B.A. in 1949 from St. Francis Xavier University, an M.S. in 1951 from the University of Illinois-Urbana-Champaign, and a Ph.D. in 1956 from Harvard University. Dr. Brush's research interests are relations between modern pollen distributions in water and surface sediments and vegetation; settling properties of pollen in water; mapping terrestrial vegetation; forest patterns; and estuarine biostratigraphy.

Eville Gorham is Regents' Professor of Ecology and Botany at the University of Minnesota, with research interests in the ecology and biogeochemistry of wetlands, global warming, and acid rain, and the chemistry of lakes and streams. He received B.S. and M.S. degrees from Dalhousie University, and a Ph.D. from the University of London. He is a member of the National Academy of Sciences, a fellow of the Royal Society of Canada, and a G. Evelyn Hutchinson Medallist of the American Society of Limnology and Oceanography.

Anthony Janetos is Senior Vice President and Chief of Program at World Resources Institute. He received an A.B. in biology in 1976 from Harvard University, an M.A. in biology in 1978 from Princeton University, and a Ph.D., also from Princeton University, in 1980. His research interests are the relationships between ecological systems and the atmosphere, and in the use of scientific information in public policy.

Arthur H. Johnson received an A.B. in geology in 1970 from Middlebury College, an M.A. in geology in 1972 from Dartmouth College, and a Ph.D. from Cornell University in soil science in 1975. He joined the University of Pennsylvania faculty in 1975 and is currently professor of geology. His research has centered on the biogeochemistry of forest ecosystems, with emphasis primarily on soil-water, soil-atmosphere, and soil-plant interactions in montane forests. Dr. Johnson is a member of the Soil Science Society of America and the American Society of Agronomy. He has participated in several National Research Council activities, including the

Committee on Long-Term Trends in Acid Deposition, the Panel on Sources and Effects of Acid Deposition, the Panel on Mechanisms of Lake Acidification, the Trends Committee, the NAS White Paper on Global Change, and the Committee on Biological Markers of Air Pollution Stress in Forests.

Daniel V. Markowitz received a B.A. in aquatic biology in 1977 from the University of California at Santa Barbara, an M.S. in marine science in 1979 from University of the Pacific, and a Ph.D. in ecology in 1987 from Kent State University. Dr. Markowitz has more than 15 years of experience in water quality evaluation, aquatic biology, project management, and environmental policy development. He is currently Associate at Malcolm Pirnie, in Akron, Ohio.

Stephen W. Pacala is professor, Department of Ecology and Evolutionary Biology, Princeton University. He is also director of graduate studies for the department associated faculty, Princeton Environmental Institute; and codirector, National Oceanic and Atmospheric Administration Carbon Modeling Center at Princeton. His research interests are plant ecology; global interactions of the biosphere, atmosphere, and hydrosphere; mathematical modeling; and community ecology. Dr. Pacala received his B.A. in 1978 from Dartmouth College and his Ph.D. in 1982 from Stanford University.

John Pastor received a B.S. in geology in 1974 from the University of Pennsylvania, an M.S. in soil science in 1977 and a Ph.D. in forestry and soil science in 1980 from the University of Wisconsin-Madison. Dr. Pastor is currently professor of biology and senior research associate, Natural Resources Research Institute at the University of Minnesota; adjunct professor, Department of Ecology and Behavioral Biology, University of Minnesota; and adjunct professor, Department of Fisheries and Wildlife, also at the University of Minnesota. His research interests are northern ecosystems, nutrient cycling, climate change, forest productivity, timber management, and landscape ecology.

Gary W. Petersen is a professor of soil and land resources in the Department of Agronomy in the College of Agricultural Sciences and codirector of the Office for Remote Sensing of Earth Resources in the Environmental Resources Research Institute at The Pennsylvania State University. His research interests have been primarily in the areas of pedology, landscape and watershed processes, land use, geographic information systems, and remote sensing. He has worked very closely with the Natural Resources Conservation Service in the areas of mapping, correlation, characterization, and interpretation. He is president of the Soil Science Society of America. Dr. Peterson received a B.S. in soils in 1961, a M.S. in soil chemistry in 1963, and a Ph.D. in soil genesis and morphology in 1965 from the University of Wisconsin.

James R. Pratt is professor of environmental science at Portland State University. His background includes degrees in biology from the University of Washington (B.A.) and Eastern Washington University (M.S.), and a Ph.D. in zoology from the Virginia Polytechnic Institute and State University. He was previously on the faculty at Murray State University and Pennsylvania State University, where he was a principal in the Environmental Resource Management Program. Dr. Pratt's research interests are microbial ecology, especially the effects of pollutants on microbial communities; and the "forgotten" microbes, the protists, including their feeding ecology, distribution, and taxonomy.

Terry Root is associate professor of natural resources at the University of Michigan. She received a B.S. in 1975 from the University of New Mexico, an M.A. in 1982 from the University of Colorado, and a Ph.D. in 1987 from Princeton University. Her research interests include ecological analyses of the distribution and abundance patterns of species on a continental scale; the physiological constraints on the distribution of wintering birds; influence of global warming on the biogeography of species; large-scale geographic examinations of the structure and composition of communities; applying quantitative information about the biogeography of species to conservation and management problems; and analyzing the ecological causes of rarity and commonness, and their effects on rare and endangered species.

Michael L. Rosenzweig is professor and former head of ecology and evolutionary biology at the University of Arizona. He is editor-in-chief of *Evolutionary Ecology*. He received an A.B. (1962) and a Ph.D. (1966) in zoology from the University of Pennsylvania. Dr. Rosenzweig uses mathematical modeling to study species diversity, habitat selection, and population interactions. With the late Robert MacArthur, he helped found modern dynamical predation theory. His work on desert rodent ecology in the United States and Israel established these systems as valuable models for the investigation of general ecological questions. His isoleg theories of habitat selection and their tests in birds and mammals have been among the first to link the study of behavior to population dynamics and community ecology. Dr. Rosenzweig's recent text, *Species Diversity in Space and Time* (Cambridge University Press, 1995) synthesizes the patterns and processes operating on diversity at scales of up to the entire planet and all of Phanerozoic time.

Milton Russell is a senior fellow at the Joint Institute for Energy and Environment, professor emeritus of economics at the University of Tennessee, and collaborating scientist, Oak Ridge National Laboratory. He was a member of the National Research Council Board on Environmental Studies and Toxicology, chairs the Committee to Assess the North American Research Strategy for Tropospheric Ozone (NARSTO) Program and is a member of the National Academy of Sciences/National Academy of Engineering Joint Committee on Cooperation in the Energy Futures of China and the United States. Dr. Russell served as Assistant Administrator for Policy at the U.S. Environmental Protection Agency from 1983 to 1987. His current research activities are concentrated on environmental policy in China and on waste management, especially cleanup of U.S. Department of Energy sites. He received his Ph.D. in economics from the University of Oklahoma in 1963.

Susan Stafford is Professor and Department Head of Applied Statistics and Research Information Management, Colorado State University. Her current interests are research information management, applied statistics, multivariate analysis and experimental design, scientific databases, GIS applications, and other data management topics. Dr. Stafford received her Ph.D. in applied statistics in 1979 from the State University of New York, College of Environmental Science and Forestry.

PROJECT DIRECTOR

David Policansky has a B.A. in biology from Stanford University and an M.S. and Ph.D., biology, from the University of Oregon. He is associate director of the Board on Environmental Studies and Toxicology at the National Research Council. His interests include genetics, evolution, and ecology, particularly the effects of fishing on fish populations, ecological risk assessment, and natural resource management. He has directed approximately 25 projects at the National Research Council on natural resources and ecological risk assessment.

Index

N

O

P

R

S

T

U

V

W